Lecture Notes on
LIMIT THEOREMS FOR MARKOV CHAIN TRANSITION PROBABILITIES

STEVEN OREY
*University of Minnesota,
Minneapolis, U.S.A.*

VAN NOSTRAND REINHOLD COMPANY
LONDON

NEW YORK   CINCINNATI   TORONTO   MELBOURNE

VAN NOSTRAND REINHOLD COMPANY
Windsor House, 46 Victoria Street, London, S.E.1

INTERNATIONAL OFFICES
New York   Cincinnati   Toronto   Melbourne

Copyright © 1971 Steven Orey

*All rights reserved. No part of this publication may be reproduced,
stored in a retrieval system, or transmitted by any means,
electronic, mechanical, photocopying, recording, or otherwise,
without the prior permission of the copyright owner.*

Library of Congress Catalog Card No. 73–141980

ISBN 0 442 06299 0

First published 1971

Printed in Great Britain by
Butler & Tanner Ltd
Frome and London

CONTENTS

Chapter 1. Limit and Decomposition Theorems

		Page
0.	Preliminaries	1
1.	Transition Probability Densities	4
2.	Existence of C-sets	7
3.	Cycles	11
4.	Invariant and Tail σ-fields	15
5.	Tail σ-fields of φ-recurrent Markov Chains	22
6.	Uniform φ-recurrence	26
7.	Invariant Measures for φ-recurrent Chains	30
8.	Normal and Anormal Chains	36
9.	φ-non-singular Chains	42

Chapter 2. Sums of Transition Probabilities 49

Chapter 3. Individual Ratio Limit Theorems

1.	Introduction and First Results	65
2.	Probabilistic Conditions for SLRP	83
3.	Necessary and Sufficient Conditions for $u_{n+1} \sim u_n$	88
4.	Conditions on the f_k implying $u_{n+1} \sim u_n$	93

Notes 99

Bibliography 103

Index 107

Introduction

In these notes some topics in the ergodic theory of discrete-time Markov chains with stationary transition probabilities and arbitrary measurable space for state space are covered. Interest will be largely confined to chains satisfying certain recurrence or irreducibility assumptions. These assumptions will ensure that the transition probabilities are well-behaved from the point of view of ergodic theory—roughly they correspond to transformations which are in some sense well-mixing. Having these rather strong hypotheses one would hope the obtain strong conclusions, and as the results will show, this is indeed possible. Most of the work is concerned with the asymptotic behaviour as n tends to infinity of the n-step transition probabilities $P^n(x, A)$ or the corresponding measures $P^n(x, \cdot)$. Many of the convergence results assert the existence of limits for all x; this contrasts with the more usual situation in ergodic theory, where results hold only almost everywhere.

The structure theorems presented in the last two sections of Chapter 1—all developments of the ideas of Doblin—give some justification for concentrating on chains satisfying a certain recurrence condition, often referred to in the literature as the recurrence condition of Harris. The theory of these chains, along with that of the much more special class of uniformly recurrent chains, is developed in Chapter 1, including the basic limit theorems. These recurrent chains always posses a unique up-to-constant factor, σ-finite, non-trivial, invariant measure π. Certain questions which are trivial when π is finite are substantial when π is infinite, and some of these are treated in Chapters 2 and 3.

First in Chapter 2 the convergence of the ratios

$$\frac{\sum_{n=0}^{N} P^n(x,A)}{\sum_{n=0}^{N} P^n(y,B)}$$

as N approaches infinity is discussed. The next question considered is the boundedness and convergence of differences of the form

$$\left| \pi(B) \sum_{n=0}^{N} P^n(x,A) - \pi(A) \sum_{n=0}^{N} P^n(y,B) \right|.$$

Chapter 3 discusses the converges of the individual ratios $P^{n+m}(x,A)/P^n(y,B)$ as n tends to infinity, where m is a fixed integer. In distinction to the situation in the first two chapters, where most questions were answered rather completely, open questions remain here. Originally most results were first discussed for the case of denumerable state space. As far as possible results are given here for general state space. However, many discrete state space results have no known extensions to general state space, and some of these are also given here. Even in the discrete state space situation there seems to be room for significant improvements upon what is known.

The problems discussed here are closely connected with the ergodic theory of a positive contraction on $L_1(S, \mathcal{B}, \lambda)$, where $(S, \mathcal{B})$ is the state space and λ a σ-finite measure on $\mathcal{B}$. If μ is a signed measure on $\mathcal{B}$ of finite total variation, let

$$\mu P(A) = \int P(x,A)\, \mu(\mathrm{d}x).$$

Now there are many finite measures λ such that λP is absolutely continuous with respect to λ, e.g. one may choose

$$\lambda = \sum_{n=0}^{\infty} 2^{-n} P^n(x,\cdot),$$

where $x \in S$. Then any signed measure μ of bounded variation which is absolutely continuous with respect to λ is transformed by P into another such signed measure μP. Now any $f \in L_1(S, \mathcal{B}, \lambda)$ is the Radon-Nikodym derivative with respect to λ of a unique signed measure μ, and if g is the Radon-Nikodym derivative of the transformed signed measure μP with respect to λ, the relation $Tf = g$ defines T as a positive contraction on $L_1(S, \mathcal{B}, \lambda)$. Conversely if λ is some σ-finite measure on $(S, \mathcal{B})$ and T is a positive contraction on $L_1(S, \mathcal{B}, \lambda)$, the adjoint T^* is a positive contraction on $L_\infty(S, \mathcal{B}, \lambda)$. For $A \in \mathcal{B}$, let I_A be the indicator function of A, i.e. $I_A(x)$ takes the value one when $x \in A$, zero otherwise, and let $P(\cdot, A) = T^*I_A$. Unfortunately $P(x,A)$ is not quite a transition probability function, the difficulty being that for fixed A $P(\cdot, A)$ is an element of $L_\infty(S, \mathcal{B}, \lambda)$ instead of an everywhere-defined function. To this extent an abstract approach via contractions on L_1 seems to be more general than a formulation starting with transition probabilities. For such an abstract approach see Foguel [1969].

Here the approach will be probabilistic. An effort has been made to make these notes essentially self-contained. However, some basic facts about discrete parameter martingales are utilized without proof. These things can of course be found in

Doob [1953], Loeve [1955], or Neveu [1956] as well as in many other books.

Historical notes are collected at the end. The references are largely restricted to works actually used, and ignore a large body of related literature. Furthermore, works which are cited will frequently contain more information than what is given here.

These notes are based on lectures given at the University of Minnesota in the spring of 1968, and at Westfield College, London, in the autumn of the same year.

CHAPTER 1

Limit and decomposition theorems

0. Preliminaries. Let S be a set and $\mathcal{B}$ a σ-field of subsets of S. A function of two arguments $P(x, B)$ is a *transition probability* function if it satisfies the following two conditions:

$P(x, \cdot)$ is a probability on $\mathcal{B}$ for every $x \in S$,

$P(\cdot, B)$ is measureable with respect to $\mathcal{B}$ for each $B \in \mathcal{B}$.

A probability measure μ on $(S, \mathcal{B})$ will be called an *initial distribution*. Let $S^\infty = S \times S \times \ldots$ and $\mathcal{B}^\infty$ the product σ-field $\mathcal{B} \times \mathcal{B} \times \ldots$. For $\omega \in S^\infty$ let $X_i(\omega)$ be the ith coordinate of ω, $i = 0, 1, \ldots$. There is a probability measure P_μ on $(S^\infty, \mathcal{B}^\infty)$ so that for any $n \geq 0$ and $B_0, B_1, \ldots, B_n$ in $\mathcal{B}$ the following relation holds:

(0.1)
$$P_\mu[X_0 \in B_0, X_1 \in B_1, \ldots, X_n \in B_n]$$
$$= \int_{B_0} \mu(dx_0) \int_{B_1} P(x_0, dx_1) \ldots \int_{B_n} P(x_{n-1}, dx_n).$$

This relation defines P_μ on cylinder sets of $\mathcal{B}^\infty$ and this measure can be uniquely extended to $\mathcal{B}^\infty$. For a justification see Doob [1953], p. 613.

The n-step transition probabilities are defined by setting $P^1(x, B) = P(x, B)$ and $P^{n+1}(x, B) = \int P^n(x, dy) P(y, B)$, where an integral sign without limits means integration over the whole space. $P^0(x, B)$ is to be equal to 1 if $x \in B$, 0 if $x \notin B$.

1

The *shift operator* θ is the maps of S^∞ into S^∞ determined by $X_n(\theta\omega) = X_{n+1}(\omega)$, $n = 0, 1, \ldots$. As usual θ^n is the nth iterate of θ, and if Y is a function on S^∞, θY is defined by $\theta Y(\omega) = Y(\theta\omega)$.

Corresponding to the probability measure P_μ is the expectation operator E_μ:

$$E_\mu Y = \int Y \, dP_\mu .$$

The probability measure ϵ_x is defined by $\epsilon_x(A) = 1(0)$ if $x \in A$ ($x \notin A$), and we write P_x, E_x for P_{ϵ_x}, E_{ϵ_x}.

Given any probability space $(G, \mathcal{F}, P)$ a measurable map Y from this space into $(S, \mathcal{B})$ is a random variable with state space $(S, \mathcal{B})$. A sequence of random variables $(X_n, n = 0, 1, \ldots)$ is a *stochastic process*. If $(\mathcal{F}_n, n = 0, 1, \ldots)$ is an increasing sequence of σ-fields, such that each $\mathcal{F}_n$ is included in $\mathcal{F}$ and $X_0, X_1, \ldots, X_n$ are all measurable with respect to $\mathcal{F}_n$ the process is *adapted* to the family $(\mathcal{F}_n)$.

Returning to the coordinate process (X_n) constructed above on the probability space $(S^\infty, \mathcal{B}^\infty, P_\mu)$, let $\mathcal{F}_n = \mathcal{B} \times \ldots \times \mathcal{B}$ (n times), to obtain an adapted process. The construction of the measure P_μ ensures that conditional probabilities can be defined so as to satisfy

$$P_\mu[X_n \in B \mid X_0 = x_0, X_1 = x_1, \ldots, X_k = x_k] = P^{n-k}(x_k, B)$$

for $n \geq 0$, $x_0 \in S$, $x_1 \in S$, $\ldots$, $x_k \in S$, $k \leq n$. The last relation is equivalent to

(0.2) $$P_\mu[X_n \in B \mid \mathcal{F}_k) = P^{n-k}(X_k, B).$$

By the indicator function of a set A is meant the function $I_A(\cdot)$ which assumes the value 1 on A and vanishes on the complement of A. Letting Y be the indicator of $[X_{n-k} \in B]$,

(0.2) becomes

(0.3) $$E_\mu[\theta^k Y \mid \mathcal{F}_k] = E_{X_k}[Y].$$

The relation (0.3) holds for any random variable measurable with respect to $\mathcal{F}_\infty$, where $\mathcal{F}_\infty = \bigcup_{k=0}^{\infty} \mathcal{F}_k$. In this form it is the most succinct expression of the *Markov property* of (X_n), and the process (X_n) is called a *Markov chain* with *stationary transition probabilities*. Since only stationary transition probabilities will be considered, this qualification will not be explicitly mentioned below. When dealing with a Markov chain one is actually dealing with a whole family of processes, one for each initial distribution μ, the various processes being connected by (0.3). Below, 'Markov process' usually refers to an individual stochastic process, but sometimes it will be used to denote the whole system of processes arising from different initial distributions. For later reference we note a consequence of (0.3), namely

(0.4) $$E_x[\theta Y] = E_x[E_x[\theta Y \mid \mathcal{F}_1]] = E_x[E_{X_1}[Y]] = \int P(x, dy) E_y[Y].$$

The following definitions and conventions will be used. Let (X_n) be a Markov chain with state space $(S, \mathcal{B})$. Then

$$Q(x, B) = P_x[\bigcap_{m=1}^{\infty} \bigcup_{n=m}^{\infty} [X_n \in B]], \qquad x \in S, \qquad B \in \mathcal{B},$$
$$L(x, B) = P_x[\bigcup_{n=1}^{\infty} [X_n \in B]], \qquad x \in S, \qquad B \in \mathcal{B}.$$

Thus $Q(x, B)$ is the probability that $X_n \in B$ for infinitely many n, given $X_1 = x$, and $L(x, B)$ is the probability that $X_n \in B$ for some positive n given $X_0 = x$.

An event of a probability space will be said to be a.s. (almost sure) if it has probability 1.

A *measure* will always mean a positive, countably additive set function. It is a feature of Doblin's approach to Markov chains that hypotheses on the Markov chain $(X_n, n = 0, 1, \ldots)$ with state space $(S, \mathcal{B})$ frequently involve an auxilliary σ-finite non-trivial measure φ on $(S, \mathcal{B})$. The letter φ will always designate such a measure. The empty set is denoted by ϕ.

Definition 0.1. Let $P(x, B)$ be a transition probability function on $(S, \mathcal{B})$. A set $E \in \mathcal{B}$ is *closed* if $E \neq \phi$ and $P(x, E) = 1$ for all $x \in E$. A closed set which does not contain two disjoint closed sets is *indecomposable*.

Let φ be a σ-finite measure on $(S, \mathcal{B})$. A Markov chain with transition probability function $P(x, B)$ is φ-*irreducible* if $L(x, E) > 0$ for all x whenever $\varphi(E) > 0$. The markov chain is φ-*recurrent* if $L(x, E) = 1$ for all x whenever $\varphi(E) > 0$.

The simplest examples to illustrate the theory to be presented are obtained by letting S be finite or denumerable, with $\mathcal{B}$ consisting of all subsets of S. In that case the transition probabilities are determined by the matrix $P(i, j) = P(i, \{j\})$, $i \in S$, $j \in S$, and counting measure is frequently a convenient choice for φ.

1. Transition probability densities. Let $P(x, B)$ be a transition probability function on $(S, \mathcal{B})$, and let φ be a σ-finite measure on $(S, \mathcal{B})$. The Radon-Nikodym theorem ensures that each $P^n(x, \cdot)$ can be decomposed into its φ-absolutely continuous and φ-singular parts

$$P^n(x, B) = \int_B p^n(x, y) \varphi(dy) + P^n_s(x, B),$$

$$x \in S, \qquad B \in \mathcal{B}, \qquad n = 1, 2, \ldots,$$

where $P^n_s(x, S^n_x) = P^n_s(x, S)$ for some $S^n_x \in \mathcal{B}$ with $\varphi(S^n_x) = 0$, and for each x $p^n(x, \ )$ is a $\mathcal{B}$-measurable function, the n-step *transition probability density*.

It will be important to be able to select well-behaved $p^n(x,y)$. In particular it is desirable to have these functions *jointly measurable* in the sense that as a function of (x, y) they are measurable with respect to $\mathcal{B} \times \mathcal{B}$. Call $\mathcal{B}$ *separable* if there exists a sub-σ-field $\mathcal{B}_\infty$ of $\mathcal{B}$ which is *countably generated*, i.e. generated by a denumerable class of sets $B_0, B_1, \ldots$, and each $B \in \mathcal{B}$ differs from a $B' \in \mathcal{B}_\infty$ only by a φ-null set.

PROPOSITION 1.1. *If $\mathcal{B}$ is separable, the transition probability densities can be chosen to be jointly measurable.*

Proof. Let $B_0, B_1, \ldots$ and $\mathcal{B}_\infty$ be as in the preceding definition of *separable*. For fixed k, $p^k(x, y)$ will be constructed as a limit of a sequence of jointly measurable functions $p(x,y)$. Let $\mathcal{B}_n$ be the finite σ-field generated by $B_0, B_1, \ldots, B_n$. Every $y \in S$ belongs to a unique atom $A_n(y)$ of $\mathcal{B}_n$. Let $p_n(x,y) = P(x, A_n(y))/\varphi(A_n(y))$, the quotient being interpreted as 0 when the denominator vanishes. Evidently $p_n(x,y)$ is jointly measurable, and consequently so is the function $p(x,y) = \liminf_{n \to \infty} p_n(x,y)$. For each fixed x, $p(x,y)$ is the limit of a sequence of difference quotients which may be expected to be the density of $P^k(x, \cdot)$ with respect to φ. A basic differentiation theorem assures that this expectation is justified, see Doob, [1953], p. 612, Theorem 2.5.

PROPOSITION 1.2. *Suppose the n-step transition probability densities are jointly measurable. Then there exist transition probability densities $p^n(x,y)$ which are jointly measurable and satisfy*

$$p^n(x,y) \geq \int P^{n-k}(x,dz)p^k(z,y)$$
(1.1)
$$\geq \int p^{n-k}(x,z)p^k(z,y)\varphi(dz), \qquad 1 \leq k \leq n, \qquad x,y \in S.$$

Proof. Let $p_0^n(x,y)$ be a version of the n-step transition probability density which is jointly measurable. For $1 \leq k \leq n$, $B \in \mathcal{B}$

$$P^n(x,B) = \int P^{n-k}(x,dz)P^k(z,B) \geq \int P^{n-k}(x,dz) \int_B p_0^k(z,y)\varphi(dy)$$

$$= \int_B \left(\int P^{n-k}(x,dz)p_0^k(z,y) \right) \varphi(dy).$$

Since $\varphi(S_x^n) = 0$ the last term is not changed by replacing B by $B - S_x^n$, and so

$$\int_B p_0^n(x,y)\varphi(dy) \geq \int_B P^{n-k}(x,dz)p_0^k(z,y)\varphi(dy)$$

showing that the $p_0^n(x,y)$ satisfy (1.1) for φ-a.e. (almost every) y. Now define $p^n(x,y)$ inductively by $p^1(x,y) = p_0^1(x,y)$,

$$p^n(x,y) = p_0^n(x,y) \vee \sup_{1 < k < n} \int P^{n-k}(x,dz)p^k(z,y), \qquad n \geq 1.$$

It follows that $p^n(x,\cdot) = p_0^n(x,\cdot)$ φ-a.e., and also that $p^n(x,y)$ is jointly measurable and the first inequalities in (1.1) hold; the second inequalities are always satisfied.

The existence of well-behaved transition probability densities is then guaranteed if $\mathcal{B}$ is separable. It will be seen below that if separability is not assumed things can occur which are ruled out by separability. However, there is a technique which allows one to extend most results from the separable situation to the general one. It depends on Proposition 1.3 below. Call a σ-field $\mathcal{B}_0$ included in $\mathcal{B}$ *admissible* if it is denumerably generated and for each $B \in \mathcal{B}_0$, $P(\cdot,B)$ is measurable with respect to $\mathcal{B}_0$.

PROPOSITION 1.3. *For any denumerable collection $\{B_n\}$ of sets of $\mathcal{B}$ there exists an admissible σ-field $\mathcal{B}_1$ such that $\{B_n\} \subseteq \mathcal{B}_1$.*

Proof. Let $\mathcal{B}_0$ be the smallest subclass of $\mathcal{B}$ which includes $\{B_n\}$ and is closed under the maps $f(A, B) = A \cup B$, $g(A) = S - A$, $f_r(A) = \{x : P(x, A) \leq r\}$ for every rational number $r \in [0,1]$. One readily sees that $\mathcal{B}_0$ is a denumerable field of sets. Let $\mathcal{B}_1$ be the σ-field generated by $\mathcal{B}_0$. To show $\mathcal{B}_1$ admissible it suffices to prove that it is included in the class $\mathcal{B}_2$ of all $B \in \mathcal{B}$ such that $P(\cdot, B)$ is measurable with respect to $\mathcal{B}_1$. Since $\mathcal{B}_0$ is closed with respect to the f_r it follows that $\mathcal{B}_0 \subseteq \mathcal{B}_2$. Now $\mathcal{B}_2$ is closed under monotone limits of sets, and so it must include along with $\mathcal{B}_0$ the σ-field generated by $\mathcal{B}_0$, that is $\mathcal{B}_1 \subseteq \mathcal{B}_2$.

2. *Existence of C-sets.* In case the state space S is denumerable, or at least contains a point s which is visited with positive probability, the situation is relatively simple, because occurrences which are separated by a visit to s are independent. In the general situation no such points s need exist, but certain sets now to be introduced partially make up for this lack.

Definition 2.1. Relative to the σ-finite measure φ on $(S, \mathcal{B})$ a set $C \in \mathcal{B}$ is called a *C-set* if $\varphi(C) > 0$ and there exists a positive integer n such that

$$\inf_{(x,y) \in C \times C} p^n(x,y) > 0.$$

THEOREM 2.1. *Let $\mathcal{B}$ be separable. Let $E \in \mathcal{B}$ satisfy $\varphi(E) > 0$ and $L(x, F) > 0$ for every $x \in E$ and every $F \subseteq E$ such that $\varphi(F) > 0$. Then E contains a C-set.*

Proof. Let φ^2 denote the product measure $\varphi \times \varphi$ on $\mathcal{B} \times \mathcal{B}$ and E^2 the set $E \times E$. For $U \subseteq S \times S$ let $U_1(x) = \{y : (x,y) \in U\}$, $U_2(y) = \{x : (x,y) \in U\}$. Let $H^{(m,n)} = \{(x,y) \in E^2 : p^m(x,y) \geq 1/n\}$, $H = \bigcup_{m=1}^{\infty} \bigcup_{n=1}^{\infty} H^{(m,n)}$. The assumption on E is easily seen to imply

(2.1) $$\varphi(H_1(x)) > 0, \qquad x \in E.$$

By Fubini's theorem $\int \varphi(H_1(x)) \varphi(dx) = \int \varphi(H_2(y)) \varphi(dy)$ and by (2.1) the integrals are positive. Using (2.1) again one obtains

$$0 < \varphi(\{y : \varphi(H_2(y)) > 0\}) = \varphi(\{y : \varphi(H_2(y)) > 0 \text{ and } \varphi(H_1(y)) > 0\}),$$

and it follows that there must exist pairs (m_1, n_1), (m_2, n_2) such that on setting $F = H^{(m_1, n_1)}$, $G = H^{(m_2, n_2)}$, one has

(2.2) $$\varphi(\{y : \varphi(F_2(y)) > 0 \text{ and } \varphi(G_1(y)) > 0\}) > 0.$$

Consider a sequence of $\mathcal{B}$-measurable finite partitions of E, each partition being a refinement of the preceding one. Each of these partitions induces a product partition on E^2. If E_α^n, E_β^n are two elements of the nth partition of E then the element $E_\alpha^n \times E_\beta^n$ of the product partition will be denoted by $E_{\alpha,\beta}^n$. For each $x \in E$ there is a unique index $i(n,x)$ such that $x \in E_{i(n,x)}^n$. We would like to assert that there exists a φ^2-null set N such that

(2.3) $$\frac{\varphi^2(F \cap E_{i(n,x),i(n,y)}^n)}{\varphi^2(E_{i(n,x),i(n,y)}^n)} \to I_F((x,y)), \qquad (x,y) \in E^2 - N,$$

and

(2.4) $$\frac{\varphi^2(G \cap E_{i(n,x),i(n,y)}^n)}{\varphi^2(E_{i(n,x),i(n,y)}^n)} \to I_G((x,y)), \qquad (x,y) \in E^2 - N,$$

the ratios in (2.3) and (2.4) being interpreted as zero when the denominator vanishes. The truth of (2.3) and (2.4) is an immediate consequence of the differentiation theorem (again see Doob [1953] p. 612, Theorem 2.5), provided only that the partitions are such that the Borel field generated by the product partitions contains F and G as members, a condition which we therefore impose on our sequence of partitions.

It follows from (2.2) and the fact that $\varphi^2(N) = 0$ that

$$\varphi(\{y : \varphi(F_2(y) - N_2(y)) > 0 \quad \text{and} \quad \varphi(G_1(y) - N_1(y)) > 0\}) > 0.$$

So there exist points x_0, y_0, z_0 with $x_0 \in F_2(y_0) - N_2(y_0)$, $z_0 \in G_1(y_0) - N_1(y_0)$. Choose n big and let $\alpha = i(n, x_0)$, $\beta = i(n, y_0)$, $\gamma = i(n, z_0)$. If n is big enough it follows from (2.3) and (2.4) that

(2.5) $$\varphi^2(F \cap E^n_{\alpha\beta}) \geq \frac{3}{4} \varphi(E^n_\alpha) \varphi(E^n_\beta),$$

and also

(2.6) $$\varphi^2(G \cap E^n_{\beta\gamma}) \geq \frac{3}{4} \varphi(E^n_\beta) \varphi(E^n_\gamma).$$

Now let

$$A = \{ x \in E^n_\alpha : \varphi(E^n_\beta \cap F_1(x)) \geq \frac{3}{4} \varphi(E^n_\beta) \},$$
$$B = \{ z \in E^n_\gamma : \varphi(E^n_\beta \cap G_2(z)) \geq \frac{3}{4} \varphi(E^n_\beta) \}.$$

Since the left members of (2.5) and (2.6) are given by

$$\int_{E^n_\alpha} \varphi(E^n_\beta \cap F_1(x)) \varphi(dx) \quad \text{and} \quad \int_{E^n_\gamma} \varphi(E^n_\beta \cap G_2(z)) \varphi(dz)$$

respectively, it is clear that $\varphi(A) > 0$ and $\varphi(B) > 0$. For $x \in A$, $z \in B$ evidently $\varphi(F_1(x) \cap G_2(z)) \geq \varphi(E^n_\beta)/2$. If also $y \in F_1(x) \cap G_2(z)$ one will have $(x,y) \in F$, $(y,z) \in G$. Recalling the definition of F and G using (1.1) one obtains

(2.5) $$p^{m_1+m_2}(x,z) \geqslant \int p^{m_1}(x,y)p^{m_2}(y,z)\varphi(dy)$$

$$\geqslant \int_{F_1(x)\cap G_2(z)} p^{m_1}(x,y)p^{m_2}(y,z)\varphi(dy)$$

$$\geqslant \frac{\varphi(E_\beta^n)}{2n_1 n_2}.$$

The last term in (2.5) is a positive constant; denote it by λ. By hypothesis $L(u, A) > 0$ for every $u \in B$ and so there exists a positive integer m and an $\eta > 0$ such that the set $C = \{x \in B : P^m(x,A) > \eta\}$ has positive φ-measure. For $x \in C$ and $y \in C$ it follows from (2.5) and (1.1) that

$$p^{m+m_1+m_2}(x,y) \geqslant \int_A P^m(x,dz) p^{m_1+m_2}(z,y) \geqslant \varphi(A)\eta\lambda > 0,$$

so that C is the desired C-set.

As an immediate consequence we have the

COROLLARY. *If $\mathcal{B}$ is separable and the Markov chain is φ-irreducible then every $E \in \mathcal{B}$ with $\varphi(E) > 0$ contains a C-set.*

Let C be a C-set with respect to φ and let $I(C)$ denote the class of positive integers n such that $\inf_{(x,y)\in C\times C} p^n(x,y) > 0$. Let $d(C)$ be the greatest common factor of $I(C)$.

PROPOSITION 2.1. *Let C be a C-set, $d(C) = d$. Then (i) $I(C)$ contains all sufficiently large multiples of d and (ii) if the chain is irreducible $d(C') = d$ for every C-set C'.*

Proof. There must exist a finite set a_i, $i = 1, 2, \ldots, k$, of members of $I(C)$ with greatest common divisor d. Let c be an integer divisible by d. Then c has the representation

(2.6) $$c = a_1 A_1 + a_2 A_2 + \ldots + a_n A_n,$$

where the A_i are integers. For each A_i there exist unique integers B_i and r_i such that $A_i = B_i a_1 + r_i$, $0 \leq r_i < a_1$, and one has the representation

(2.7) $$c = a_1(A_1 + a_2 B_2 + \ldots + a_n B_n) + a_2 r_2 + \ldots + a_n r_n,$$

where all coefficients except that of a_1 are non-negative. Comparing the expression for c/a_1 obtained by dividing (2.6) by a_1 with the coefficient of a_1 in (2.7) reveals that this coefficient exceeds $(c/a_1) - (a_2 + a_3 + \ldots + a_n)$, and the last quantity is positive for all big enough c. The truth of (ii) is evident. Since $I(C)$ is closed under addition (i) follows.

3. *Cycles.* The term *cycle* will be defined below, and the main business of this section is to prove that under mild assumptions the state space of a Markov chain permits a decomposition yielding a cycle, and to show that this decomposition is essentially unique.

The following proposition will be useful.

PROPOSITION 3.1. *Let $(Q, \mathcal{F}, P)$ be a probability space, $(\mathcal{F}_n)$, $n = 1, 2, \ldots$ an increasing sequence of σ-fields, each $\mathcal{F}_n \subseteq \mathcal{F}$ and let $\mathcal{F}_\infty = \bigcup_{n=1}^\infty \mathcal{F}_n$. Let $A_i \in \mathcal{F}_\infty$, $i = 1, 2, \ldots$. Then*

(i) $P\left[\bigcap_{i=n}^\infty A_i \mid \mathcal{F}_n\right] \to I_{\bigcup_{m=1}^\infty \bigcap_{i=m}^\infty A_i}$ *as* $n \to \infty$.

(ii) $P\left[\bigcup_{i=n}^\infty A_i \mid \mathcal{F}_n\right] \to I_{\bigcap_{m=1}^\infty \bigcup_{i=m}^\infty A_i}$ *as* $n \to \infty$.

Proof. Since (i) and (ii) have similar proofs only the argument for (ii) will be given. For $k \leqslant n$

$$P\left[\bigcup_{i=k}^{\infty} A_i \mid \mathcal{F}_n\right] \geqslant P\left[\bigcup_{i=n}^{\infty} A_i \mid \mathcal{F}_n\right] \geqslant P\left[\bigcap_{m=1}^{\infty} \bigcup_{i=m}^{\infty} A_i \mid \mathcal{F}_n\right].$$

Applying the martingale convergence theorem to the extreme members gives

$$I_{\left[\bigcup_{i=k}^{\infty} A_i\right]} \geqslant \limsup_n P\left[\bigcup_{i=n}^{\infty} A_i \mid \mathcal{F}_n\right]$$
$$\geqslant \liminf_n P\left[\bigcup_{i=n}^{\infty} A_i \mid \mathcal{F}_n\right] \geqslant I_{\left[\bigcap_{m=1}^{\infty} \bigcup_{i=m}^{\infty} A_i\right]}.$$

As $k \to \infty$ the term on the extreme left approaches that on the extreme right, and the proposition is proved.

The following definition presents some important concepts introduced by Doblin. Recall the notations $Q(x, E)$ and $L(x, E)$ defined after formula (0.4).

Definition 3.1. Let $P(x, B)$ be a transition probability function on $(S, \mathcal{B})$. The set $E \in \mathcal{B}$ is *inessential* if $Q(x, E) = 0$ for all $x \in S$; otherwise E is *essential*. An essential set which is the union of denumerably many inessential sets is *improperly essential*; otherwise it is *properly essential*.

Remark. The terminology differs slightly from that introduced by Doblin: he uses 'absolutely essential' in place of 'properly essential'.

It will turn out that for our purposes a set which is not properly essential is small; in many cases, e.g. when the chain is φ-recurrent, all essential sets are properly essential.

PROPOSITION 3.2. *Let E be closed. Let $E^0 = \{x : L(x,E) = 0\}$. Then $S - (E \cup E^0)$ is not properly essential.*

Proof. It suffices to show that the set $F_m = \{x : L(x,E) \geq 1/m, x \notin E\}$ is inessential for $m = 1, 2, \ldots$. Now

$$P_x\left[\bigcup_{i=n}^{\infty}[X_{1+i} \in E] \mid \mathcal{F}_n\right] = P_x\left[\theta_n\left[\bigcup_{i=0}^{\infty}[X_{1+i} \in E] \mid \mathcal{F}_n\right]\right] = L(X_n, E)$$

and by Proposition 3.1 the term on the left tends to the indicator function of $\bigcap_{m=1}^{\infty}\bigcup_{i=m}^{\infty}[X_{1+i} \in E]$ P_x-a.s. Thus P_x-a.s., if $X_n \in F_m$ for infinitely many n, then $X_n \in E$ for arbitrarily large n; but since E is closed this implies $X_n \in E$ for all sufficiently big n, and this contradicts $X_n \in F_m$ for infinitely many n.

Definition 3.2. A sequence $(C_1, C_2, \ldots, C_k)$ of k disjoint sets is a *cycle* (of length k) if each $C_j \in \mathcal{B}$ and

$$P(x, C_{j+1}) = 1, \qquad x \in C_j, \qquad 1 \leq j \leq k-1,$$

$$P(x, C_1) = 1, \qquad x \in C_k.$$

THEOREM 3.1. *Suppose the Markov chain on $(S,\mathcal{B})$ is φ-irreducible. Then there exists a cycle $(C_1, C_2, \ldots, C_d)$, such that the following conditions hold:*

(i) *The set $S - \bigcup_{i=1}^{d} C_i$ is a φ-null set and it is not properly essential.*

(ii) *If $C_1^1, C_2^1, \ldots, C_{d^1}^1)$ is a cycle, then d^1 divides d, and for each i, $i = 1, 2, \ldots, d^1$, C_i^1 differs from a union of d/d^1 members of $\{C : 1 \leq j \leq d\}$ only by a φ-null set which is not properly essential.*

Proof. Assume first that $\mathcal{B}$ is separable. By Theorem 2.1 there exists a C-set C; let $d = d(C)$ be as in Proposition 2.1. Let $\tilde{C}_j$ be

the set of all x such that $P^{nd-j}(x,C) > 0$ for some $n \geq 1$. The
assumption of φ-irreducibility implies every x is in some $\tilde{C}_j$.
However, the $\tilde{C}_j$ need not be disjoint.

For $1 < k < d$ call $E \in \mathcal{B}$ *k-accessible* if there exists a subset
A of C of positive φ-measure such that $P_x\left[\bigcup_{n=0}^{\infty} [X_{nd+k} \in E]\right] > 0$ for
all $x \in A$; if E is k-accessible for some k it is *accessible*.
Evidently C is d-accessible. On the other hand C cannot be
k-accessible for any k less than d, for otherwise one sees easily,
with the help of (1.1), that the class $I(C)$ would contain
integers equal to k modulo d, contradicting the definition of d. This
implies that $\tilde{C}_i \cap \tilde{C}_j$ is not accessible for $i \neq j$, and hence the same
is true for $F = \cup[\tilde{C}_i \cap \tilde{C}_j]$, the union extending over all pairs (i,j)
such that $0 < i < j \leq d$. Note that $S-F$ is closed. Let $C_i = \tilde{C}_i - F$,
$i = 1, 2, \ldots, d$. Now $F = S - \bigcup_{i=1}^{d} C_i$. Since F is not accessible
$\varphi(F) = 0$ by φ-irreducibility. Now apply Proposition 3.2 with $S-F$
in place of E. By φ-irreducibility E^0 must be empty, and the
conclusion becomes that F is not properly essential.

Very similar, straightforward arguments give the uniqueness
assertion (ii).

Finally the assumption that $\mathcal{B}$ is separable will be removed.
For an admissible σ-field $\mathcal{B}$, the argument above applies to give an
essentially unique cycle $(C_1, C_2, \ldots, C_{d_1})$. If $\mathcal{B}_2$ is an admissible
σ-field which includes $\mathcal{B}$ and $(C_1^1, C_2^1, \ldots, C_{d_2}^1)$ a cycle for $(S, \mathcal{B}_2)$,
then by the uniqueness part of the theorem either $d_2 = d_1$ and the
two cycles are essentially identical, or d_1 divides d_2 and the second
cycle is a refinement of the first. It will suffice to show that there
exists an admissible σ-field $\mathcal{B}_1$ such that for any admissible $\mathcal{B}_2$
including $\mathcal{B}_1$ the corresponding d_1 and d_2 agree. For assume on the
contrary that there exists an increasing sequence $\mathcal{B}_i$ of admissible

σ-fields with associated integers d_i strictly increasing. By Proposition 1.3 there exists an admissible σ-field $\mathcal{B}^*$ including all the sets belonging to the cycles associated with the $\mathcal{B}_i$, $i = 1, 2, \ldots$. Then $\mathcal{B}^*$ contains cycles of length d_i, $i = 1, 2, \ldots$, which contradicts (ii).

When Theorem 3.1 applies, the chain will be called *aperiodic* if $d = 1$, *periodic* if $d > 1$; d is the *period* associated with the chain.

4. *Invariant and tail σ-fields.* Let $(X_n, n = 0, 1, \ldots)$ be a stochastic process defined on a probability space $(\Omega, \mathcal{F}, P)$, the random variables X_n assuming values in a measurable space $(S, \mathcal{B})$. Let $\mathcal{F}^n$ be the σ-field generated by $\{X_m : m \geq n\}$, $n = 0, 1, \ldots$, and let $\mathcal{F}^\infty = \bigcap_{n=0}^{\infty} \mathcal{F}^n$. A random variable Y is a *tail random variable* if there exists a sequence (f_n) of $\mathcal{B}^\infty$-measurable functions on S^∞ such that

(4.1) $\qquad Y = f_n(X_n, X_{n+1}, \ldots), \qquad n = 0, 1, \ldots.$

In case Y is the indicator of an event, the event is a *tail event*. If in (4.1) f_n can be chosen independently of n, so that

(4.2) $\qquad Y = f(X_n, X_{n+1}, \ldots), \qquad n = 0, 1, \ldots,$

Y is called *invariant*. Again if $Y = I_A$, and Y is invariant the event A is called an *invariant event*. The class of all tail events is a σ-field, the *tail σ-field*, which coincides with $\mathcal{F}^\infty$. Similarly the class of all invariant events constitutes a σ-field: the *invariant σ-field*.

The above notions did not involve the probability measure P. On identifying events and random variables which agree P-a.s. one obtains the notions of *P-tail random variable (event)*, and

P-invariant random variables (event), and *P-tail σ-field, P-invariant σ-field*. A two element σ-field will be called *trivial*.

Let $N = \{0, 1, \ldots\}$ and let $\mathcal{N}$ consist of all subsets of N. A process $((X_n, T_n) : n = 0, 1, \ldots)$ with X_n a random variable taking values in $(S, \mathcal{B})$ and T_n a random variable taking values in $(N, \mathcal{N})$, $n = 0, 1, \ldots$, is called a *space-time process (associated with $(X_n, n = 0, 1, \ldots))$* if $T_{n+1} = T_n + 1$, $n = 0, 1, \ldots$.

PROPOSITION 4.1. *Let $((X_n, T_n), n = 0, 1, \ldots)$ be a space-time process such that $T_0 = k$ for some k. Then an event A is a tail event for $(X_n, n = 0, 1, \ldots$ if and only if A is an invariant event for $((X_n, T_n), n = 0, 1, \ldots)$.*

Proof. Let Y be the indicator of an event A and suppose Y satisfies (4.1) for a sequence of $\mathcal{B}^\infty$-measurable functions (f_n). Let $f((x_0, n_0), (x_1, n_1), \ldots) = f_{n_0 - k}(x_0, x_1, \ldots)$ to obtain $Y = f((X_n, T_n), (X_{n+1}, T_{n+1}), \ldots)$, $n = 0, 1, \ldots$. Conversely, if the last equality obtains, setting $f_n(x_0, x_1, \ldots) = f((x_0, n+k), (x_1, n+k+1), \ldots)$ implies (4.1)

Let $P(x, B)$ be the transition probability function for a Markov process on $(S, \mathcal{B})$. A real valued $\mathcal{B}$-measurable function on S is *harmonic* if $f(x) = \int f(y) P(x, dy)$. If $X_0, X_1, \ldots$ is a Markov chain with transition probability functions $P(x, B)$, then any associated space-time process $(X_0, T_0), (X_1, T_1), \ldots$ is a Markov chain with transition probabilities $\tilde{P}(\cdot, \cdot)$ determined by

$$\tilde{P}((x, n), E \times \{n+1\}) = P(x, E), \qquad x \in S, \qquad E \in \mathcal{B}, \qquad n \in N.$$

A function g on $S \times N$ harmonic for the space-time process will be called *space-time harmonic*. So if g is space-time harmonic

(4.3) $$g(x, n) = \int P(x, dy)\, g(y, n+1).$$

Let $P(x, B)$ be a transition probability function on $(S, \mathcal{B})$ and associate with it a Markov chain $(X_n, n = 0, 1, \ldots)$ as in Section 0, so that $X_n(\omega) = \omega_n$, where $(\omega_n) \in S^\infty$. The space-time chain $(Z_n, n = 0, 1, \ldots)$, $Z_n = (X_n, T_n)$ is constructed similarly: $Z_n(\tilde{\omega}) = \tilde{\omega}_n$, where $\tilde{\omega} = (\tilde{\omega}_n) \in (S \times N)^\infty$, $\tilde{\omega}_n = (\omega_n, t_n)$. It is true that not all $\tilde{\omega} \in (S \times N)^\infty$ are allowed, since $t_{n+1} = t_n + 1$ is required, but this is immaterial.

For Markov chains the shift operator θ was defined in Section 0. For a random variable H, let $\theta^k H$ be the random variable defined by $\theta^k H(\omega) = H(\theta^k \omega)$. Returning to the Markov chain of the previous paragraph, note that H is invariant if and only if $\theta H = H$. It should be emphasized that different initial probability distributions μ result in different probability measures P_μ, and the notion of a P_μ-tail event or P_μ-invariant event depend on μ; however, the concepts of tail events and invariant events involve only the product space $(S^\infty, \mathcal{B}^\infty)$ and the shift θ.

Consider the map from random variables H on $(S^\infty, \mathcal{B}^\infty)$ to measurable functions h on $(S, \mathcal{B})$ given by $h_H(x) = E_x[H]$. It has the following properties:

PROPOSITION 4.2. *(i) If H is bounded and invariant h_H is harmonic and bounded, and h_H vanishes if and only if $P_x[H = 0] = 1$ for all x.*

(ii) If h is bounded and harmonic, the invariant random variable $H = \liminf_{n \to \infty} h(X_n)$ satisfies $h_H = h$.

Proof. Let H be bounded and invariant. Then $\theta^n H = H$, $n = 1, 2, \ldots$ and if $h(x) = E_x[H]$, (0.3) implies

(4.4) $\qquad h(X_n) = E_{X_n}[H] = E[\theta_n H \,|\, \mathcal{F}_n] = E[H \,|\, \mathcal{F}_n].$

Specializing to $n = 1$ and applying $E_x[\cdot]$ to the extreme members gives
$$E_x[h(X_1)] = E_x[H] = h(x),$$
which expresses that h is harmonic. This implies that $(h(X_n), n = 0, 1, \ldots)$ is a martingale. The identity between the first and last members of (4.4) also makes the martingale property apparent, and immediately implies $E_x h(X_n) = E_x[H] = h(x)$. By the martingale convergence theorem $h(X_n) \to H$ P_x-a.s. for any $x \in S$, so that h vanishes identically if and only if $P_x[H = 0] = 1$ for all x. This proves both (i) and (ii).

COROLLARY. *The following two conditions are equivalent:*
(i) All bounded harmonic functions are constant.
(ii) The σ-field of strictly P_μ-invariant events is trivial for every initial probability μ.

Proof. By Proposition 4.2 implies that all bounded invariant random variables are constant P_μ-a.s. so the σ-field of P_μ-invariant sets is trivial. For the opposite direction suppose there exists a non-constant bounded harmonic function h; say $h(x) \neq h(y)$. By Proposition 4.2 $h(z) = E_z[H]$ for some invariant H. Let $\mu = 1/2(\epsilon_x + \epsilon_y)$. Then $E_\mu[H|X_0 = x] = h(x) \neq h(y) = E_\mu[H|X_0 = y]$ and so H is not constant P_μ-a.s.

For the next result recall the notation introduced at the beginning of this section.

THEOREM 4.1. *The stochastic process $(X_n, n = 0, 1, \ldots)$ satisfies*

(4.5) $$\lim_{n\to\infty} \sup_{A \in \mathcal{F}^n} |P(A \cap B) - P(A)P(B)| = 0,$$

for every $B \in \mathcal{F}$ if and only if the P-tail σ-field is trivial.

Proof. Assume (4.5) holds and let $B \in \mathcal{F}^\infty$. Consider (4.5) for $A = B$ and obtain $P(B) = P(B^2)$, proving that $\mathcal{F}^\infty$ modulo P-null sets is trivial. For $B \in \mathcal{F}$, $A \in \mathcal{F}^n$ one has:

$$|P(A \cap B) - P(A)P(B)| = |\int_A [P(B|\mathcal{F}^n) - P(B)]\,dP|$$
$$\leq \int |P(B|\mathcal{F}^n) - P(B)|\,dP,$$

and the backward martingale convergence theorem (Doob [1953], p. 382 Theorem 4.2) asserts that $P(B|\mathcal{F}^n)$ converges to $P(B|\mathcal{F}^\infty)$-a.s. If $\mathcal{F}^\infty$ modulo P-null sets is trivial $P(B|\mathcal{F}^\infty) = P(B)$-a.s. and the theorem is proved.

Definition 4.1. If μ is a signed measure on $(S, \mathcal{B})$ let $||\mu||$ be the total variation. If $P(x, B)$ are transition probabilities μP is the measure defined by

$$\mu P(A) = \int P(x, A)\,\mu(dx), \text{ for every } A \in \mathcal{B}.$$

In the next proposition we again consider a Markov chain $(X_n,\ n = 0, 1, \ldots)$ on $(S, \mathcal{B})$ with transition probabilities $P(x, B)$; let $\widetilde{P}(z, E)$ be the corresponding space-time transition probability function, and $(\widetilde{X}_n,\ n = 0, 1, \ldots)$ the space-time chain, and denote by $\widetilde{P}_\eta[\ ]$ the probability measure corresponding to the space-time chain with initial probability η and transition probabilities $\widetilde{P}(\cdot, \cdot)$.

PROPOSITION 4.3. *The following conditions are equivalent:*
 (i) *For every initial space-time probability distribution η such that $\eta(S \times \{k\}) = 1$ for some k, the $\widetilde{P}_\eta$-invariant σ-field of $(\widetilde{X}_n)$ is trivial.*
 (ii) *For every initial probability distribution μ the P_μ-tail σ-field of (X_n) is trivial.*

(iii) *For all probability measures μ and ν on $\mathcal{B}$,*
$$\lim_{n \to \infty} ||\mu P^n - \nu P^n|| \to 0.$$

(iv) *The only bounded space-time harmonic functions are constants.*

Proof. It will be shown that (i) $\Rightarrow$ (ii) $\Rightarrow$ (iii) $\Rightarrow$ (iv) $\Rightarrow$ (i). To show that (i) $\Rightarrow$ (ii) apply Proposition 4.1.

Next assume (ii). Let x and y be two distinct points of S, and let $\mu = 1/2(\epsilon_x + \epsilon_y)$. Let $A = [X_0 = x]$. By Theorem 4.1 relation (4.5) applies to give:

(4.6) $\quad |P_\mu [A \cap [X_n \in E]] - P_\mu(A) P_\mu [X_n \in E]|$
$$= \frac{1}{4} | P^n(x,E) - P^n(y,E) | \to 0,$$

as $n \to \infty$ uniformly for $E \in \mathcal{B}$.

Still assume (ii) and let μ and ν be two probability measures on $\mathcal{B}$ and let H_n be the positive set of the Hahn decomposition of $(\mu-\nu)P^n$.

Then $(\mu P^n - \nu P^n)(H_n) = \iint (P^n(x,H_n) - P^n(y,H_n))\mu(dx)\nu(dy)$,

and the last term approaches zero by (4.6) and the bounded convergence theorem. Since the same argument applies to the negative set of the Hahn decomposition (iii) follows.

Assume (iii). Let h be a space-time harmonic function bounded in absolute value by $M < \infty$. For any $x, y \in S$ and any positive integer k, iterating (4.3) k times gives

$$h(x,n) - h(y,n) = \int h(z, n+k)(P^k(x, dz) - P^k(y, dz))$$
$$\leqslant M ||P^k(x, \cdot) - P^k(y, \cdot)||,$$

and by assumption the last term tends to zero as $k \to \infty$. Hence h is a function of n alone, and therefore, as (4.3) makes evident, h must in fact be constant, and (iv) is proved.

The implication from (iv) to (i) follows by the corollary to Proposition 4.2 and Proposition 4.1.

It should be noted that the condition that the P_x-invariant σ-field is trivial for every $x \in S$ is weaker than (ii).

The P_μ-tail σ-field of a Markov chain can be much larger than the P_μ-invariant σ-field, as shown by the following example.

Example 4.1. Let the state space S be denumberable, and let $\mathcal{B}$ consist of all subsets of S. Let the states be labelled (n,m), $n = 0, 1, \ldots, 1 \leqslant m \leqslant 2^n$. The non-zero entries in the transition probability matrix are the following: $P_{(n,m)(n,m+1)} = 1$, $1 \leqslant m \leqslant 2^n$; $P_{(n,2^n)(n+1,1)} = P_{(n,2^n)(n+2,1)} = \frac{1}{2}$. For $(n,m) \in S$ the corresponding n will be referred to as the *column index* of the state. Let μ be the unit mass at $(1,1)$. Let the random variables $N_0, N_1, \ldots$ be the distinct successive column indexes assumed by the Markov chain (X_i), and let $U_k = (N_k - N_{k-1})$, $k = 1, 2, \ldots$. Thus the U_k assume the values 1 or 2. Note that the number $K+1$ of distinct column indexes the chain has assumed by time n, as well as the values $N_0, N_1, \ldots, N_K$ of these indexes can be calculated from a knowledge of X_n. Hence each of the events $[U_k = 1]$, $[U_k = 2]$, $k = 1, 2, \ldots$ is equal modulo P_μ-null sets to a tail event, and the P_μ-tail σ-field is not atomic. On the other hand there are no bounded harmonic functions other than constants, so the P_μ-invariant σ-field is trivial. The same argument also works if μ is concentrated on an arbitrary point (n_0, m_0). Finally one sees easily that the conclusion P_μ-invariant σ-field trivial but P_μ-tail σ-field not atomic is true for any initial probability distribution μ.

5. *Tail σ-fields of φ-recurrent Markov chains.* It will be shown that φ-recurrent Markov chains have P_μ-tail σ-fields which are finite, and the atoms will be identified. When the chain is aperiodic the P_μ-tail σ-field is trivial, and most questions can easily be reduced to this case.

Consider a Markov chain (X_n) with state space $(S, \mathcal{B})$. For $B \in \mathcal{B}$ let $\Lambda(B) = \bigcap_{m=1}^{\infty} \bigcup_{i=m}^{\infty} [X_i \in B]$, and let $\Lambda^c(B)$ be the complement of $\Lambda(B)$.

PROPOSITION 5.1. *(i) Let $E \in \mathcal{B}$, $F \in \mathcal{B}$ be such that $\inf_{x \in E} L(x, F) > 0$. Then for any initial probability measure μ, $\Lambda(E) \subseteq \Lambda(F)$ P_μ-a.s.; therefore $P_\mu[\Lambda(E)] \leq P_\mu[\Lambda(F)]$.*

(ii) Let $E \in \mathcal{B}$, $F \in \mathcal{B}$ be such that $\sup_{x \in E} Q(x, F) < 1$. Then for any initial probability measure μ, $\Lambda(E) \subseteq \Lambda^c(F)$ P-a.s.; therefore $P_\mu[\Lambda(E) \cap \Lambda(F)] = 0$.

Proof. Let $A_i = [X_{i+1} \in E]$. Then $L(X_n, F) = P[\bigcup_{i=n}^{\infty} A_i | \mathcal{F}_n] \to I_{\Lambda(F)}$ as $n \to \infty$, P_μ-a.s. by Proposition 3.1. If the hypothesis of (i) holds one therefore has:

$$\Lambda(E) \subseteq [\limsup_{n \to \infty} L(X_n, F) > 0] \subseteq [\lim_{n \to \infty} L(X_n, F) = 1] = \Lambda(F)$$

P_μ-a.s. and (i) is proved. Next, $Q(X_n, F) = P[\bigcap_{m=0}^{\infty} \bigcup_{i=m}^{\infty} A_i | \mathcal{F}_n] \to I_{\Lambda(F)}$ as $n \to \infty$, P_μ-a.s. by the martingale convergence theorem. Hence if the hypothesis of (ii) holds:

$$\Lambda(E) \subseteq [\liminf_{n \to \infty} Q(X_n, F) < 1] \subseteq [\lim_{n \to \infty} Q(X_n, F) = 0] \to I_{\Lambda^c(F)}$$

P_μ-a.s. and (ii) is proved.

COROLLARY. *Let (X_n) be φ-recurrent. Then $\varphi(A) > 0$ implies $Q(x, A) = 1$ for all $x \in S$ and every bounded harmonic function is consta*

Proof. Suppose $\varphi(A) > 0$. Then $L(x,A) = 1$ for all $x \in S$ and so the hypothesis of Proposition 5.1 (i) holds with $E = S$, $F = A$, and the conclusion of (i) with $\mu = \epsilon_x$ gives $1 = Q(x,S) \leqslant Q(x,A) \leqslant 1$, proving the first assertion of the corollary. Let h be a bounded harmonic function; by Proposition 4.2 there is a bounded invariant H such that $h(X_n) \to H$ as $n \to \infty$ P_x-a.s. for every $x \in S$. Let a_0 be the least upper bound of all a such that $\varphi\{x : h(x) \geqslant a\} > 0$. If $a < a_0$ $Q(x,\{y : h(y) \geqslant a\}) = 1$ and so $H \geqslant a$ P_x-a,s, for all x; if $a > a_0$, $Q(x,\{y : h(y) < a\}) = 1$ and so $H \leqslant a$ P_x-a.s. So evidently $P_x[H = a_0] = 1$ for all x, and consequently $h(x) = E_x[H] = a_0$.

THEOREM 5.1. *Let (X_n) be a Markov chain with state space $(S, \mathcal{B})$ and suppose that the chain is φ-recurrent. If the chain is aperiodic the P_μ-tail σ-field is trivial for every initial probability distribution μ. If the chain is periodic with period d and $(C_1, C_2, ..., C_d)$ is a cycle the P_μ-tail σ-field is atomic, the atoms consisting of the equivalence classes modulo P_μ-null sets corresponding to the events $E_i = \bigcup_{m=0}^{\infty} \bigcap_{n=m}^{\infty} [X_{nd} \in C_i]$ such that $P_\mu[E_i] > 0$, $1 \leqslant i \leqslant d$.*

Remark. When $d > 0$ there are at most d atoms corresponding to $E_1, E_2, ..., E_d$; however, if $P_\mu[E_i] = 0$ for some i there will be fewer than d atoms.

Proof. Consider first the aperiodic case. By Proposition 4.3 what has to be shown is that any bounded space-time harmonic function is constant. Let h be such a function. Let $B_{r,n} = \{x \in S : h(x,n) < r\}$ for r rational, $n = 0,1,...$. This denumerable class of sets is contained in an admissible σ-field $\mathcal{B}_0$ and h is measurable with respect to $\mathcal{B}_0 \times \mathcal{N}$. In order to prove h constant we may consider in place of the original chain the one with state space

$(S, \mathcal{B}_0)$, where $P(x,B)$ is unchanged, except that B is restricted to $\mathcal{B}_0$. Putting it another way, we may assume that $\mathcal{B}$ is separable, since the general case can be reduced to this case, and so this assumption will now be made. Define h' by $h'(x,n) = h(x,n+1)$ and note that h' is also a space-time harmonic function. It will be shown that $h = h'$, so that $h(x,n) = h^*(x)$. Then h^* is bounded and harmonic; hence h^* is constant by the Corollary to Proposition 5.1.

Suppose $z_0 = (x_0, n_0)$, $h(z_0) \neq h'(z_0)$. Let (Z_n) be the space-time chain (X_n, T_n). The bounded martingales $(h(Z_n))$, $(h'(Z_n))$ converge P_{z_0}-a.s. to limits H and H', respectively. The assumption on z_0 implies $P_{z_0}[H \neq H'] > 0$. Assume $P_{z_0}[H < H'] > 0$; the other case can be treated in exactly the same way. Then there exist numbers a and b with $a < b$ such that $P_{z_0}[H < a, H' > b] = \delta > 0$. Let $A = \{z : h(z) < a\}$, $B = \{z : h'(z) > b\}$. Then $P_{z_0}[\bigcup_{m=0}^{\infty} \bigcap_{n=m}^{\infty} [Z_n \in A \cap B)] \geq \delta$. Let $g(z) = P_z[\bigcap_{n=0}^{\infty} [Z_n \in A \cap B]]$. It follows from Proposition 3.1 and the Markov property of (Z_n) that $g(Z_n)$ converges P_{z_0}-a.s. to the indicator of $\bigcup_{m=0}^{\infty} \bigcap_{n=m}^{\infty} [Z_n \in A \cap B]$. Thus $g(z)$ can be made arbitrarily close to 1 by appropriate choice of z. By the Corollary to Theorem 2.1 C-sets exist; let C be such a set. Since we are in the aperiodic situation there exist, by Proposition 2.1, a positive integer m and $\epsilon > 0$ such that $p^m(x,y) > \epsilon$ and $p^{m+1}(x,y) > \epsilon$, for all $x \in C$ and all $y \in C$. By the Corollary to Proposition 5.1, $Q(z_0, C) = 1$, so that Z_n enters C infinitely often P_{z_0}-a.s. Since $g(Z_n)$ converges to 1 with positive P_{z_0}-probability it follows that with positive P_{z_0}-probability both $Z_n \in C$ and $g(Z_n)$ close to 1 will occur, for the same n. So there must be a point $z_1 = (x_1, n_1)$ with $x_1 \in C$ such that $1 - g(z_1) < \epsilon \varphi(C)/4$. Let $C' = C \cap \{x : (x, n_1 + m) \notin A \cap B\}$. Note that:

$\epsilon\varphi(C') \leq P_{x_1}[X_m \in C'] \leq P_{z_1}[Z_m \notin A \cap B] \leq 1 - g(z_1) \leq \epsilon\varphi(C)/4.$

Similarly, if $C'' = C \cap \{x : (x, n_1+m+1) \notin A \cap B\}$, then $\varphi(C'') \leq \epsilon\varphi(C)/4$. So C contains a point x which does not belong to $C' \cup C''$, and for this x $(x, n_1+m) \in A \cap B$, $(x, n_1+m+1) \in A \cap B$. But this is impossible, since $(x,n) \in B$ says that $h(x,n+1) > b$, and this implies $h(x,n+1) \notin A$. Thus $h \neq h'$, has had to be shown.

The periodic case is easily reduced to the aperiodic one by using Theorem 3.1.

COROLLARY. *Let (X_n) and $(S, \mathcal{B})$ be as in Theorem 5.1. If the chain is aperiodic, then for any two probability measures μ and ν $||(\mu-\nu)P^n|| \to 0$ as $n \to \infty$. In case the chain has period d,*

$$||d^{-1} \sum_{k=0}^{d-1} (\mu-\nu)P^{n+k}|| \to 0 \text{ as } n \to \infty \text{ for any pair of probability}$$

measures μ and ν.

Proof. In the aperiodic case this follows immediately from Theorem 5.1 and Proposition 4.3. Simple considerations reduce the periodic case to the aperiodic one.

Remark 5.1. Many Markov chains satisfy a condition related to but weaker than φ-recurrence: if $\varphi(B) > 0$ then $Q(x,B) = 1$ for φ-a.e. x. Consider such a chain, and let Y be the indicator of an invariant event. For every initial probability distribution μ, if $h(x) = E_x[Y]$, $h[X_n]$ is a martingale converging to Y P_μ-a.s. Hence if $0 < a < 1$, the sets $\{x : h(x) > a\}$, $\{x : h(x) < a\}$ cannot both have positive φ-measure. Indeed it follows that $h(y) = 1$ φ-a.e. or $h(y) = 0$ φ-a.e. Thus if μ is absolutely continuous with respect to φ, $P_\mu[A] = 0$ or $P_\mu[A] = 1$ for each invariant A, so that for such μ the P_μ-invariant σ-field is trivial. In case μ is a positive σ-finite measure it is sometimes convenient to introduce P_μ by the definition $P_\mu(A) = \int P_x[A]\mu(dx)$. In this case it is still true if μ is absolutely continuous with respect to φ the P_μ-invariant σ-field is trivial.

6. *Uniform φ-recurrence.* In this section a condition still stronger than φ-recurremce will be introduced. When this condition holds the conclusions of the previous sections can be strengthened. Furthermore these results will prove useful in the study of chains which do not themselves satisfy the condition.

It will be useful to have available the notation

(6.1) $\quad _B P^m(x, A) = P_x[X_m \in A, \ X_i \notin B, \ 1 \leqslant i < m],$

$\qquad x \in S, \qquad B \in \mathcal{B}, \qquad A \in \mathcal{B}, \qquad m = 1, 2, \ldots .$

Definition 6.1. The Markov chain (X_n) on $(S, \mathcal{B})$ is *uniformly φ-recurrent* if $\sum_{m=1}^{n} {}_A P^m(x, A) \to 1$ uniformly in x as $n \to \infty$ whenever $\varphi(A) > 0$.

The next proposition serves as a key lemma for the important Theorem 7.1.

PROPOSITION 6.1. *Consider a chain on $(S, \mathcal{B})$ which is uniformly φ-recurrent. If the chain is aperiodic, there exist $a < \infty$ and $\rho < 1$ such that*

(6.2) $\qquad\qquad ||(\mu-\nu)P^n|| \leqslant a\rho^n ||\mu-\nu||$

for any two probability measures μ and ν on $(S, \mathcal{B})$. If the chain is periodic there exist $a < \infty$, $\rho < 1$ such that

$$||\frac{1}{n} \sum_{k=1}^{n} (\mu-\nu)P^k|| \leqslant a\rho^n ||\mu-\nu|| .$$

LIMIT AND DECOMPOSITION THEOREMS

Proof. Assume the chain is aperiodic. Let μ, ν be probability measures on $(S, \mathcal{B})$. It is easy to see that $||(\mu-\nu)P^n||$ is non-increasing as a function of n. It will be shown that there exists an n_1 and $\rho_1 < 1$, independent of μ and ν such that

(6.3) $$||(\mu-\nu)P^{n_1}|| \leqslant \rho_1 ||\mu-\nu||.$$

Iterating gives $||(\mu-\nu)P^{kn_1}|| < \rho_1^k ||\mu-\nu||$, $k = 1, 2, \ldots$ and this together with the monotonicity of $||(\mu-\nu)P^n||$ is easily seen to imply the desired conclusion. For any μ, ν there exist mutually singular measures μ' and ν' such that $\mu-\nu = \mu'-\nu'$, $||\mu'|| = ||\nu'|| = ||\mu-\nu||/2$. It follows from

$$||(\mu-\nu)P^n|| = ||(\mu'-\nu')P^n||$$
$$= (||\mu'||)^{-1} ||\iint ((\epsilon_x - \epsilon_y)P^n)\mu'(dx)\nu'(dy)|| \leqslant$$
$$\leqslant (||\mu'||)^{-1} \iint ||(\epsilon_x - \epsilon_y)P^n|| \mu'(dx)\nu'(dy)$$

that it suffices to prove (6.3) for the case $\mu = \epsilon_x$, $\nu = \epsilon_y$, $x \neq y$.

Let $x_0 \in S$, $\delta > 0$. By the Corollary to Theorem 5.1, as n approaches infinity $||(\epsilon_{x_0} - \epsilon_z)P^n||$ tends to 0 for each z, and so there exists $B \in \mathcal{B}$ with $\varphi(B) > 0$, and n_0 such that $||(\epsilon_{x_0} - \epsilon_z)P^{n_0}|| < \delta/4$ holds for all $z \in B$. By hypothesis there exists n_1 such that $\sum_{k=1}^{n_1} {}_B P^k(y, B) > 1 - (\delta/4)$ holds for all y. For $n \geqslant n_0 \vee n_1$, and any $A \in \mathcal{B}$, $y \in S$,

(6.4) $$(P^n(x_0, A) - P^n(y, A)) - \sum_{k=1}^{n_1} \int_B (P^n(x_0, A) - P^{n-k}(z, A))_B P^k(y, dz) \leqslant$$
$$2 \sum_{k=n_1+1}^{\infty} {}_B P^k(y, B) \leqslant \frac{\delta}{2}.$$

Note now that

$$
(6.5) \quad \sum_{k=1}^{n_1} \int_B (P^n(x_0, A) - P^{n-k}(z, A))_B P^k(y, dz)
$$

$$
= \sum_{k=1}^{n_1} \int_B (P^{n-k}(x_0, A) - P^{n-k}(z, A))_B P^k(y, dz)
$$

$$
+ \sum_{k=1}^{n_1} \int_B (P^n(x_0, A) - P^{n-k}(x_0, A))_B P^k(y, dz).
$$

It follows from the definition of B that the first term on the right tends to 0 as n approaches infinity, the convergence being uniform in A. By the Corollary to Theorem 5.1, $||P^n(x_0, \cdot) - P^{n-k}(x_0, \cdot)|| = ||\epsilon_{x_0} - \epsilon_{x_0} P^k) P^{n-k}||$ tends to zero for each fixed k as n tends to infinity, and hence the last term in (6.5) tends to zero uniformly in A and y as n goes to infinity. So there exists an n_2 such that for $n > n_2$ the second term in (6.4) is bounded in absolute value by $\delta/2$. Thus as n approaches infinity the first term of (6.4) tends to zero uniformly in A and y. Since

$$|P^n(x, A) - P^n(y, A)| \leq |P^n(x, A) - P^n(x_0, A)| + |P^n(x_0, A) - P^n(y, A)|,$$

the left side tends to 0 uniformly in x, y, and A as n approaches infinity. Hence there certainly exists $\rho_1 < 1$ and n_1 so that (6.3) holds for $\mu = \epsilon_x$, $\nu = \epsilon_y$, as had to be shown. Routine considerations take care of the periodic case.

For $A \in \mathcal{B}$, $x \in S$, $B \in \mathcal{B}$ introduce the notation:

$$
(6.6) \quad P_A(x, B) = \sum_{k=1}^{\infty} {}_A P^k(x, B).
$$

It should be noted that $P_A(x,B)$ is the expected number of visits (X_n) pays to B after time 0 up to and including the first time A is entered, given $X_0 = x$. Observe also that if $Q(x,A) = 1$ for all $x \in A$, $P_A(x,B)$ defines a transition probability function on $(A, \mathcal{B}_A)$, where $\mathcal{B}_A$ is the class of all subsets of A which belong to $\mathcal{B}$. The intuitive significance of this transition probability is the following. If the Markov chain on $(S, \mathcal{B})$ with transition probabilities $P(x,B)$ starts at x, $x \in A$, and if T_1, T_2, ... are the successive return times to A, then X_0, X_{T_1}, X_{T_2}, ... is a Markov chain with state space $(A, \mathcal{B}_A)$ and transition probability $P_A(x,B)$, $B \in \mathcal{B}_A$. This chain is called the *process on A*.

Definition 6.1. Consider a Markov chain on $(S, \mathcal{B})$. A set $D \in \mathcal{B}$ is a *D-set* if the process on D is uniformly φ-recurrent.

PROPOSITION 6.1. *Suppose a Markov chain on $(S, \mathcal{B})$ satisfies the following condition: for every $A \in \mathcal{B}$ such that $\varphi(A) > 0$ there exist $n > 0$ and $\epsilon > 0$ such that $\sum_{k=1}^{n} {}_A P^k(x,A) > \epsilon$ for all $x \in S$. Then the chain is uniformly φ-recurrent.*

Proof. Indeed it is evident that $\sum_{k=1}^{jn} {}_A P^k(x,A) \geq 1 - (1-\epsilon)^j$, $j = 1, 2, \ldots$.

THEOREM 6.2. *Let (X_n) be a Markov chain on $(S, \mathcal{B})$ which is φ-recurrent, and suppose $\mathcal{B}$ is separable. Then there exist $D_n \in \mathcal{B}$, $n = 1, 2, \ldots$ such that $D_1 \subset D_2 \ldots$ and $\bigcup_{n=1}^{\infty} D_n = S$ and each D_n is a D-set.*

Proof. By Theorem 2.1 there exists a C-set C. Say then that $p^{n_0}(x,y) > \epsilon > 0$ for all $x \in C$, $y \in C$. Let $n(x)$ be the least n such that $P_x[\bigcup_{i=1}^{n}[X_i \in C]] \geq 1/n$, and set $D_n = \{x : n(x) \leq n_0 + n\}$. Note that $n(x)$ is a $\mathcal{B}$-measurable function, so $D_n \in \mathcal{B}$. Since $D_n \uparrow S$ is immediate it remains only to show that each D_n is a D-set. To see this let $A \subseteq D_n$, $\varphi(A) > 0$. For $1 > \delta > 0$ there exists a $B \in \mathcal{B}$ and $n_1 > 0$ such that $B \subseteq C$, $\varphi(B) > 0$, and $P_z[\bigcup_{k=1}^{n_1}[X_k \in A]] > \delta$ for all $z \in B$. Starting at any $x \in D_n$ the probability of reaching C at some positive time not exceeding n is at least $1/n$; from any point in C one enters B in exactly n_0 steps with probability at least $\epsilon\varphi(B)$, and from any point in B one can get to A in n_1 or fewer steps with probability greater than δ. So $x \in D_n$ implies

(6.7) $$P_x[\bigcup_{k=1}^{n+n_0+n_1}[X_k \in A]] > \frac{1}{n} \cdot \epsilon\varphi(B) \cdot \delta,$$

and since the left side is majorized by the probability that the process on D_n enters A by time $n+n_0+n_1$, given that it starts at x, Proposition 6.1 applies and the theorem is proved.

7. *Invariant measures for φ-recurrent chains.* Existence and uniqueness of invariant measures will be discussed; in certain cases the measures $P^n(x,\cdot)$ will be seen to converge to an invariant probability measure.

Definition 7.1. A σ-finite measure π on $(S,\mathcal{B})$ is *invariant* for the transition probability function $P(x,\mathcal{B})$ if

(7.1) $$\pi(A) = \int P(x,A)\pi(dx), \qquad x \in S, \qquad A \in \mathcal{B}.$$

If also $\pi(S) = 1$, π is an invariant probability measure.

THEOREM 7.1. *(i) If a φ-recurrent chain has an invariant probability measure π, then $||\mu P^n - \pi|| \to 0$ a $n \to \infty$ for every initial probability measure μ, provided the chain is aperiodic; in case the* chain has period d, $||d^{-1} \sum_{k=0}^{d-1} \mu P^{n+k} - \pi|| \to 0$.

(ii) A φ-recurrent chain has at most one invariant probability measure.

*(iii) A uniformly φ-recurrent chain has an invariant probability measure π. There exist a finite constant a and $\rho < 1$ such that for every initial probability measure μ, $||(\mu-\pi)P^n|| \leq a\rho^n$ in the aperiodic case; in the periodic case the conclusion is
$||d^{-1} \sum_{k=0}^{d-1} \mu P^{n+k} - \pi|| \leq a\rho^n$, where d is the period.*

Proof. The truth of (i) follows from the Corollary to Theorem 5.1. Evidently (i) implies (ii).

To prove (iii) assume the chain is uniformly φ-recurrent. Assume also that the chain is aperiodic. Let μ be an initial probability distribution. Then

$$||\mu P^n - \mu P^{n+m}|| = ||(\mu - \mu P^m)P^n||,$$

and by Proposition 6.1 the last term tends to 0 as n approaches infinity, uniformly in m. Therefore $\mu P^n(A)$ converges to a limit $\pi(A)$ uniformly in A, $A \in \mathcal{B}$, and π will be a probability measure on $(S, \mathcal{B})$. Passing to the limit in $P^{n+1}(x, A) = \int P(y, A) P^n(x, dy)$ shows that π satisfies (7.1). The final assertion of (iii) in the aperiodic case also follows from Proposition 6.1. The periodic case can be handled similarly.

The next theorem settles the invariant measure question for φ-recurrent chains.

THEOREM 7.2. *For every φ-recurrent chain on $(S, \mathcal{B})$ there exists a non-trivial σ-finite measure π such that the following conditions hold.*

(i) π *is invariant.*

(ii) *If π' is σ-finite and invariant, π' is a constant multiple of π.*

(iii) φ *is absolutely continuous with respect to π.*

Proof. Suppose first that $\mathcal{B}$ is separable. Let $A \in \mathcal{B}$ be such that $\varphi(A) > 0$ and the process on A has an invariant probability measure π_A. According to Theorem 7.1 this will be the case if A is a D-set and Theorem 6.2 guarantees the existence of D-sets. Define a measure π on $(S, \mathcal{B})$ by

$$(7.2) \qquad \pi(E) = \int_A P_A(x, E) \pi_A(dx),$$

where $P_A(x, E)$ was defined by (6.6). Observe that if $E \subseteq A$, $\pi(E) = \pi_A(E)$.

If the Markov chain is started with initial distribution π_A, then $\pi(E)$ is just the expected number of times
$[X_m \in E, X_1 \notin A, X_2 \notin A, \ldots, X_{m-1} \notin A, m > 0]$, so evidently if $S_{n,m} = \{x : P^m(x, A) > 1/n\}$, then $\pi(S_{n,m}) < \infty$. Since $S = \bigcup_{n,m} S_{n,m}$, π is σ-finite.

To show invariance write:

$$(7.3) \quad \int \pi(dy) P(y, E) = \int_A \pi_A(dy) P(y, E) + \int_{S-A} \int_A \pi_A(dx) P_A(x, dy) \cdot P(y, E)$$

$$= \int_A \pi_A(dx) [P(x, E) + \int_{S-A} P_A(x, dy) P(y, E)]$$

$$= \int_A \pi_A(dx) P_A(x, E) = \pi(E).$$

Next let π be an arbitrary σ-finite invariant measure, and suppose $A \in \mathcal{B}$, $\pi(A) = 1$. Let π_A be the restriction of π to subsets of A which belong to $\mathcal{B}$. It is to be shown that (7.2) holds for all $E \in \mathcal{B}$ such that $E \subseteq A$. Note that this will imply that π_A is the invariant probability measure for the process on A. It will first be shown that for any $E \in \mathcal{B}$,

$$\text{(7.4)} \quad \pi(E) = \sum_{k=1}^{n} \int_{A} {}_A P^k(x,E) \pi(dx)$$
$$+ \int_{S-A} {}_A P^n(x,E) \pi(dx), \quad n = 1, 2, \ldots,$$

where ${}_A P^k(x,E)$ was defined in (6.1). This is proved by induction. For $n = 1$, (7.4) follows from the definition of invariant measure, since ${}_A P^1(x,E) = P(x,E)$. Suppose now that (7.4) holds for some n. The last term in (7.4) can be written thus:

$$\int_{S-A} {}_A P^n(x,E) \pi(dx) = \int_{x \in S-A} {}_A P^n(x,E) \int_{y \in S} \pi(dy) P(y,dx)$$
$$= \int_{y \in A} \pi(dy) \int_{x \in S-A} {}_A P^n(x,E) P(y,dx) +$$
$$+ \int_{y \in S-A} \pi(dy) \int_{x \in S-A} {}_A P^n(x,E) P(y,dx)$$
$$= \int_{A} {}_A P^{n+1}(y,E) \pi(dy) + \int_{S-A} {}_A P^{n+1}(y,E) \pi(dy),$$

and on using this relation in (7.4) one obtains the desired identity for $n+1$. Letting n tend to infinity in (7.4) implies

$$\text{(7.5)} \quad \pi(E) \geqslant \int_{A} P_A(x,E) \pi_A(dx) \quad E \in \mathcal{B}.$$

If $F \in \mathcal{B}$, $F \subseteq A$, (7.5) must hold with $E = F$ and also with $E = A-F$. Also

$$\pi(F) + \pi(A-F) = \pi(A) = 1 = \int_{A} P_A(x,F) \pi_A(dx) + \int_{A} P_A(x,A-F) \pi_A(dx),$$

where the last equality holds because the two terms of the last member represent the probability that the process on A with initial probability distribution π_A enters F, respectively $A-F$, at time 1. It follows that $\pi(F) = \int_{A} P_A(x,F) \pi_A(dx)$, as was to be shown. Hence if π and π' are two σ-finite invariant measures and $A \in \mathcal{B}$ such that $0 < \pi(A) < \infty$, $0 < \pi'(A) < \infty$, then the measures $[\pi(A)]^{-1} \cdot \pi(\cdot)$ and $[\pi'(A)]^{-1} \pi'(\cdot)$ must both agree on subsets of A with the unique invariant probability measure for the process on A, and it is easily deduced that these measures must in fact agree on all of $\mathcal{B}$, so that (ii) holds.

Let π be a σ-finite invariant measure. Then $\pi(E) = 0$ implies that $P^n(\cdot, E) = 0$ for all n π-a.e. However, $\varphi(E) > 0$ implies $L(x,E) = 1$ and hence $\sum_{n=1}^{\infty} P^n(x,E) > 0$ for all x, so that (iii) follows.

In the arguments above the assumption that $\mathcal{B}$ is separable was used in the proof of (i). If $\mathcal{B}$ is not separable let $\mathcal{B}_0$ be an admissible countably generated σ-field, π_0 an invariant measure on $(S, \mathcal{B}_0)$. For every admissible $\mathcal{B}_1$ such that $\mathcal{B}_0 \subseteq \mathcal{B}_1$, there is a unique invariant measure π_1 agreeing with π_0 on $\mathcal{B}_0$. Given any $A \in \mathcal{B}$ there exists an admissible σ-field $\mathcal{B}_1$ such that $\mathcal{B}_0 \subseteq \mathcal{B}_1$, $A \in \mathcal{B}_1$, and hence a corresponding π_1. Set $\pi(A) = \pi_1(A)$. Note that although A does not uniquely determine $\mathcal{B}_1$, it does uniquely determine $\pi_1(A)$. Now $\pi(A)$ is defined for every $A \in \mathcal{B}$, and π is easily seen to be a σ-finite invariant measure.

COROLLARY. *Let π be a non-trivial σ-finite invariant measure for a φ-recurrent chain on $(S, \mathcal{B})$. Then the chain is π-recurrent and for every $B \in \mathcal{B}$ the following are equivalent: (i) $\pi(B) > 0$, (ii) $L(x, B) > 0$ for all $x \in S$, (iii) $Q(x, B) = 1$ for all $x \in B$.*

Proof. It is evident that (iii) implies (ii) and that (ii) implies (i). Since for any closed set E, the measure agreeing with π on subsets of E and vanishing on $S - E$ is invariant, the uniqueness assertion (ii) of Theorem 7.2 shows that $\pi(S - E) = 0$. Suppose $\pi(B) > 0$. If $B^0 = \{x : L(x\ B) = 0\} \neq \phi$, $B^0 \cap (S - B)$ is closed. This leads to a contradiction with the previous remarks: so $B^0 = \phi$. Hence there exists $A \in \mathcal{B}$ with $\varphi(A) > 0$, and $L(x, B)$ bounded away from 0. By Proposition 5.1 and its corollary, $Q(y, B) \geqslant Q(y, A) = 1$ for all y.

The next theorem gives information about the asymptotic behaviour of $P^n(x,A)$ for φ-recurrent chains with infinite invariant measure.

THEOREM 7.3. *Let π be a σ-finite invariant measure for a Markov chain on $(S, \mathcal{B})$, and suppose that the chain is π-recurrent and $\pi(S) = \infty$. For every $\epsilon > 0$ and every $x \in S$*

$$\frac{P^n(x,E)}{\pi(E) + \epsilon} \to 0 \; uniformly \; in \; E \in \mathcal{B}, \qquad as \; n \to \infty.$$

Proof. Assume that the chain is aperiodic; the general case is easily reduced to this situation. Suppose the conclusion of the theorem is not true. Then there exists a strictly increasing sequence of integers (n_k), a sequence of elements (E_k) of $\mathcal{B}$, an $x_0 \in S$ and $\delta > 0$, $\epsilon > 0$ such that

(7.6) $$\frac{P^{n_k}(x_0, E_k)}{\pi(E_k) + \epsilon} \geqslant \delta, \qquad k = 1, 2, \ldots.$$

It follows from the Corollary to Theorem 5.1, the fact that $\pi(S) = \infty$, and Egorov's theorem that there exists a positive integer k and $B \in \mathcal{B}$ such that $\pi(B) > 1/\delta$, and $|P^{n_k}(x_0, E_k) - P^{n_k}(y, E_k)| \leqslant \epsilon \cdot \delta/2$ for all $y \in \mathcal{B}$. Therefore, remebering (7.6),

$$\pi(E_k) = \int P^{n_k}(x, E_k)\pi(dx) \geqslant \int_B P^{n_k}(x, E_k)\pi(dx)$$
$$\geqslant \pi(B)(P^{n_k}(x_0, E_k) - \frac{\epsilon\delta}{2})$$
$$\geqslant \pi(B)(\delta(\pi(E_k) + \epsilon) - \frac{\epsilon\delta}{2}),$$

and thus

$$\pi(B) \leqslant \frac{\pi(E_k)}{\delta(\pi(E_k) + \epsilon/2)} \leqslant \frac{1}{\delta}$$

contradicting the choice of B.

COROLLARY. *Let π be a σ-finite invariant measure for a chain on $(S,\mathcal{B})$, and suppose that the chain is π-recurrent, $\pi(S) = \infty$. For every $E \in \mathcal{B}$ with $\varphi(E) < \infty$, $P^n(x,E) \to 0$ as $n \to \infty$ for each $x \in S$.*

Proof. Immediate from the Theorem.

8. *Normal and anormal chains.* On the basis of Definition 8.1 below, indecomposable properly essential chains are divided into two classes, called *normal* and *anormal*, respectively. It was understood by Doblin, and will be made clear below that the terminology can be taken seriously: mild conditions, satisfied by most indecomposable and properly essential chains that one is liable to encounter naturally, guarantee normality.

Definition 8.1. A Markov chain on $(S,\mathcal{B})$ is *normal* if S is indecomposable and properly essential and there exists a closed subset F of S which contains no improperly essential subsets; such an F is called a *final set*. If S is indecomposable and properly essential but contains no final set the chain in *anormal*.

A closed set H such that the process on H is φ-recurrent for some φ is a final set. It follows from Theorem 8.2 that the converse is also true.

Because of its frequent occurence below it is convenient to introduce some notation by another definition.

Definition 8.2. Consider a Markov chain on $(S,\mathcal{B})$. For $E \in \mathcal{B}$, let $E^0 = \{x : L(x,E) = 0\}$, $E^\infty = \{x : Q(x,E) = 1\}$.

The following simple proposition will be useful below.

PROPOSITION 8.1. *Suppose S is indecomposable and properly essential. For $E \in \mathcal{B}$ the following conditions are equivalent. (i) E is properly essential. (ii) $E^0 = \phi$ (iii) $E^\infty \neq \phi$.*

Proof. Let $E_n = \{x : Q(x,E) \leq 1-n^{-1}, L(x,E) > n^{-1}\}$, $n = 1, 2, \ldots$. Then $S = E^0 \cup E^\infty \cup \bigcup_{n=1}^\infty E_n$. It follows from Proposition 5.1 that E_n is not properly essential, $n = 1, 2, \ldots$. Since S is properly essential, either E^0 or E^∞ must be too. If both E^0 and E^∞ are non-void, E^∞ and $(S-E) \cap E^0$ are two disjoint closed sets; this contradicts the indecomposability of S, so that $E^0 = 0$ or $E^\infty = 0$. Thus the equivalence of (ii) and (iii) follows. Assume (iii) holds. Then E^∞ is properly essential. To conclude that (i) holds suppose otherwise: $E = \bigcup_{n=1}^\infty F_n$, each F_n being inessential. For $x \in E^\infty$ one has

$$L(x\ E) = \lim_{n \to \infty} L(x, \bigcup_{k=1}^n F_k) = 1.$$

Let $G_n = \{x \in E^\infty : L(x, \bigcup_{k=1}^n F_k) > n^{-1}\}$. Then $E^\infty = \bigcup_{k=1}^\infty G_n$ and for some n G_n must be properly essential. So for some $x \in S$, $Q(x, G_n) > 0$ and hence by Proposition 5.1 $Q(x, \bigcup_{n=1}^n F_k) > 0$, which is not compatible with $F_1, F_2, \ldots, F_n$ being inessential. Finally, if $E^\infty = \phi$, $E = \bigcup_{n=1}^\infty (E \cap E_n) \cup (E \cap E_0)$ displays E as a denumberable union of sets which are not properly essential, so (i) implies (iii).

COROLLARY. *If S is indecomposable and A is closed, $S-A$ is not properly essential.*

Proof. Since $A \subseteq (S-A)^0$ the result follows from Proposition 8.1.

The dot in notations such as $A \dot\cup B$ or $\dot\cup A_i$ is used to indicate that the sets whose union is being taken are pairwise disjoint.

THEOREM 8.1. *Consider a φ-irreducible chain on $(S,\mathcal{B})$. Suppose that $\mathcal{B}$ is separable. Then S can be partitioned*

$$S = H \cup I,$$

where I is not properly essential, and either $H = \phi$ or H is closed and the process on H is φ-recurrent.

Proof. By Theorem 2.1 there exists a C-set C. Let $H = C^\infty$. Then H is closed or void. In either case $I = S-H$ is not properly essential: in the first case this follows from the preceding corollary, in the second it is proved by simple arguments like those above.

It is to be shown that if H is not void the process on H is φ-recurrent. Since H is closed, φ-irreducibility implies $\varphi(I) = 0$, hence $\varphi(H) > 0$. Let $E \subseteq H$, $\varphi(E) > 0$. It follows from φ-irreducibility that there exists an $F \subseteq C$, $\varphi(F) > 0$ with $L(x,E)$ bounded away from 0 as x ranges over F. By Proposition 5.1 $Q(x,E) \geqslant Q(x,F)$ for all $x \in S$. On the other hand, since C is a C-set $L(x,F)$ is bounded away from 0 for x ranging over C, and again by Proposition 5.1 $Q(x,F) \geqslant Q(x,C)$ for all $x \in S$. Since $Q(x,C) = 1$ for $x \in H$, it follows that $Q(x,E) = 1$, proving H φ-recurrent.

COROLLARY. *Consider a φ-irreducible chain on $(S,\mathcal{B})$. If S is properly essential there exists a non-trivial σ-finite invariant measure π. This measure is unique up to constant factor and φ is absolutely continuous with respect to π.*

Proof. If $\mathcal{B}$ is separable consider the decomposition $S = H \cup I$ of the Theorem. Since S is properly essential $H \neq \phi$, and the process on H has a σ-finite invariant measure π_H. This measure can be extended to a measure π on $(S,\mathcal{B})$ in exactly one way so that $\pi(I) = 0$. Clearly π is invariant. The asserted absolute continuity follows from Theorem 7.2. Finally, the case when $\mathcal{B}$ is not separable can be reduced to the separable case in the same manner this was done in the last paragraph of the proof in Theorem 1.2.

The next theorem describes the situation under the assumption that S is indecomposable and properly essential.

THEOREM 8.2. *Consider a chain on $(S,\mathcal{B})$ and suppose S is indecomposable and properly essential.*

(i) *If the chain is normal, $S = H \cup I$, where I is not properly essential, and H is a closed set such that the process on H is φ-recurrent for any φ which assigns positive mass to every closed set; such φ exist.*

(ii) *If $\mathcal{B}$ is separable the chain is normal.*

(iii) *In any case there exists a non-trivial σ-finite invariant measure φ, and this measure is unique up to multiplicative constant, and there exists a cycle $(C_1, C_2, ..., C_d)$ satisfying the conditions of Theorem 3.1.*

Proof. Let φ be a σ-finite measure on $(S,\mathcal{B})$ such that $\varphi(A) > 0$ for each closed A. Such measures exist: for any $x \in S$, $\sum_{n=1}^{\infty} 2^{-n} P^n(x,\cdot)$ satisfies this condition by virtue of indecomposability.

Next, a closed set K will be obtained such that whenever $A \subseteq K$, and $\varphi(A) > 0$ then A is properly essential. Consider all subclasses of $\mathcal{B}$ whose members are pairwise disjoint inessential sets of positive φ-measure. By Zorn's lemma there exists a class $\mathcal{Q}$ in this collection which is maximal under $\subsetneq$. Note that $\mathcal{Q}$ is an at most denumerable class of inessential sets and so $A = \cup \mathcal{Q}$ is not properly essential. Proposition 8.1 shows that A^0 is not void. It follows that $K = A^0 \cap (S-A)$ is closed. In view of the maximality of $\mathcal{Q}$, K can contain no inessential set of positive φ-measure, and hence, since any improperly essential set of positive φ-measure contains an inessential set of positive φ-measure, any subset of K of positive φ-measure is properly essential.

Assume now that the chain is normal, and let F be a final set. Set $H = F \cap K$. Since S is indecomposable H is non-void, and evidently H is closed. By the Corollary to Proposition 8.1, $I = S-H$ is not properly essential. Let $B \subseteq H$, $\varphi(B) > 0$. Since $B \subseteq K$, B is properly essential, and so by Proposition 8.1 $B^\infty \neq \phi$, and so evidently $H \cap B^\infty \neq \phi$. Applying the Corollary of Proposition 8.1 to the process on H shows $H-B^\infty$ is not properly essential, and since $H-B^\infty \subseteq F$, it follows that $H-B^\infty$ is inessential. So $L(x, B^\infty) = 1$ for all $x \in H-B^\infty$, hence for all $x \in H$. This proves (i).

For (ii) assume $\mathcal{B}$ separable. Again consider K. Let $B \subseteq K$, $\varphi(B) > 0$, so that by construction of K, B is properly essential. According to Proposition 8.1 $L(x,B) > 0$ for all x, and this implies that the process on K is φ irreducible. By Theorem 8.1, which can be applied to the process on K because K is properly essential, K contains a closed set H such that the process on H is φ-recurrent. Again the Corollary to Proposition 8.1 implies that $I = S-H$ is not properly essential.

When the chain is normal, assertion (iii) is easily reduced, by means of (i) to the corresponding assertion for φ-recurrent chains, which has been shown to be true in Theorem 7.2, and Theorem 3.1. By (ii), if $\mathcal{B}$ is separable the chain is normal. Thus it only remains to treat the non-separable case by the use of admissible σ-fields in the usual manner, e.g. see the proofs of Theorem 3.1 and Theorem 7.2.

Example of an anormal chain. The state space is $(S, \mathcal{B})$ with $S = \{x : 0 \leqslant x < \infty\}$ and $\mathcal{B}$ the class of all subsets B of S such that either B or $S-B$ is at most denumerable. Let (a_n) be a sequence of numbers, $0 < a_n < 1$, $n = 0, 1, \ldots$ such that $\prod_{n=0}^{\infty} a_n > 0$. Let η be the measure on $(S, \mathcal{B})$ such that $\eta(B) = 0$ if B is at most denumerable, $\eta(B) = 1$ otherwise. Let

$$P(x, E) = a_{[x]} I_E(x+1) + (1-a_{[x]})\eta(E), \qquad x \in S, \qquad E \in \mathcal{B},$$

where $[x]$ is the greatest integer not exceeding x. One verifies readily that $P(x, E)$ is a transition probability on $(S, \mathcal{B})$. Observe (i) $E \in \mathcal{B}$ is essential if and only if there exists an x such that $\{x+n : n = 0, 1, \ldots\} \subseteq E$, (ii) $E \in \mathcal{B}$ is closed if and only if E is not finite or denumerable, and whenever $x \in E$ also $x+1 \in E$. It follows from (ii) that S is indecomposable. To show that S is properly essential suppose $S = \bigcup_{i=1}^{\infty} S_i$, $S_i \in \mathcal{B}$, $i = 1, 2, \ldots$. Then for some S_n, $(S-S_n)$ is denumerable or finite. Then evidently for some $x \{x+m : m = 0, 1, \ldots\} \subseteq S_n$ and S_n is essential. Since any closed set F will contain a subset of the form $\{x+n : n = 0, 1, \ldots\}$ which is improperly essential, the chain is not normal.

THEOREM 8.3. *Consider a Markov chain on $(S, \mathcal{B})$ with S indecomposable and properly essential. The chain is normal if and only if the function*

$$g(x) = \sup\{Q(x, E) : E \in \mathcal{B}, E \text{ improperly essential}\}$$

is $\mathcal{B}$-measurable.

Proof. It follows from Proposition 8.1 that for every improperly essential E, $Q(x, E) < 1$ for all x. Suppose (E_n) is a sequence of improperly essential sets such that $Q(x, E_n) \uparrow g(x)$. Then, since the denumerable union of improperly essential sets is still improperly essential $1 > Q(x, \bigcup_{n=1}^{\infty} E_n) \geq g(x)$. Let $A_n = \{x : g(x) \leq 1-n^{-1}\}$. If g

is $\mathcal{B}$-measurable each $A_n \in \mathcal{B}$. In that case, since S is properly essential, some A_n is properly essential. By Proposition 8.1 A_n^∞ is not void, hence it is closed. For $x \in A_n^\infty$, $Q(x, A_n) = 1$ and it follows by means of Proposition 5.1 that $Q(x, E) = 0$ for every improperly essential E. In particular if $E \subseteq A_n^\infty$, $E \in \mathcal{B}$, and E is improperly essential then $Q(x, E) = 0$ for all $x \in E$ and E is inessential. Thus A_n^∞ is a final set.

Conversely, if F is a final set $g(x) = Q(x, S-F)$ is evident, hence g is $\mathcal{B}$-measurable.

9. *φ-non-singular chains.* A basic decomposition result of Doblin will be given, which sometimes allows reduction to situations discussed earlier.

The following simple proposition will motivate the subsequent definition.

PROPOSITION 9.1. *Consider a Markov chain on $(S, \mathcal{B})$ and a σ-finite measure φ on $\mathcal{B}$. The following conditions are equivalent. (i) For every closed set E, $\varphi(E) > 0$. (ii) For every $x \in S$ there exists an n such that $P^n(x, \cdot)$ is not φ-singular.*

Proof. It is evident that (ii) implies (i). Suppose (ii) fails, so that for some x and every n, $P^n(x, \cdot)$ is singular with respect to φ. Then there exists a φ-null set F such that $L(x, S-F) = 0$. Hence $(S-F)^0 \neq \phi$, and $F \cap (S-F)^0$ is a closed set of φ-measure zero. So (i) fails.

Definition 9.1. If either of the two equivalent conditions in Proposition 9.1 holds the chain is called *φ-non-singular.*

Definition 9.2. An indecomposable set not strictly contained in any indecomposable set is *maximal indecomposable.*

LIMIT AND DECOMPOSITION THEOREMS

The next three propositions are easy; they will be useful below. Note that for any $F \in \mathcal{B}$, F^0 is either void or closed, and if F is closed $F \cap F^0 = \phi$.

PROPOSITION 9.2. *If E is indecomposable, $(E^0)^0$ is a maximal indecomposable set containing E.*

Proof. Observe $E^0 \cap (E^0)^0 = \phi$, so $L(x, E) > 0$ for all $x \in (E^0)^0$. Hence for any closed subset C of $(E^0)^0$, $C \cap E$ is non-void and therefore closed. So the existence of two disjoint closed subsets C and D of $(E^0)^0$ would imply the existence of disjoint closed subsets $E \cap C$ and $E \cap D$ of E, contradicting the indecomposability of E. To prove the maximality, suppose $(E^0)^0$ is properly contained in a closed set F. For $x \in F - (E^0)^0$, $L(x, E^0) > 0$, and so $E^0 \cap F$ is not void, and hence it is closed. Then $F \cap E^0$ and $F \cap E$ are disjoint closed subsets of F, and F is not indecomposable.

PROPOSITION 9.3. *Distinct maximal indecomposable sets are disjoint.*

Proof. Let E and F be distinct maximal indecomposable sets. Then the closed sets $E \cup F$ is not indecomposable, and so it contains two disjoint closed sets C and D. Since E is indecomposable, either C or D must intersect E vacuously. Say $C \cap E = \phi$. Then $C \subseteq F - E \subseteq F$, and since F is indecomposable $F \cap E$, which is closed or void, must in fact be void.

PROPOSITION 9.4. *Let E be closed, F a closed subset of E. If $E - F$ contains no closed sets, $E - F$ is not properly essential.*

Proof. Suppose $E - F$ contains no closed set. Then $(F^0 - F) \cap E$ must be void, since otherwise it would be a closed subset of $E - F$. Therefore $E - F = \bigcup_{n=1}^{\infty} G_n$, where $G_n = \{x \in E - F : L(x, F) \geq 1/n\}$. It

is to be shown that G_n is not properly essential, $n = 1, 2, \ldots$.
Suppose otherwise: $Q(x, G_n) > 0$ for some n, and some $x \in S$. Using
Proposition 5.1 and the notation introduced there one obtains
$0 < P_x[\Lambda(G_n)] = P_x[\Lambda(G_n) \cap \Lambda(F)]$. However the last term must
equal 0 since F is closed and $G_n \subseteq S - F$. This contradiction
completes the proof.

The next two propositions constitute the key steps in the proof
of Theorem 9.1.

PROPOSITION 9.5. *Consider a φ-non-singular chain on $(S, \mathcal{B})$.
Let $E \in \mathcal{B}$, $\varphi(E) < \infty$, E closed, and suppose E contains no
indecomposable subsets. Then there exist disjoint closed subsets
E_1 and E_2 of E such that $E - (E_1 \cup E_2)$ contains no closed subset
and $\varphi(E_2) \leqslant \varphi(E_1) \leqslant 2/3\ \varphi(E)$.*

Proof. Let $\mathcal{E}$ be the class of all pairs (E_1, E_2) such that E_1
and E_2 are disjoint closed subsets of E, $\varphi(E_2) \leqslant \varphi(E_1)$, and
$E - (E_1 \cup E_2)$ contains no closed subset. $\mathcal{E}$ is not void. For by the
assumptions on E, there exists a closed subset F of E such that
$E - F$ contains a closed subset. Letting G be a closed subset of
$E - F$ of maximal measure, that is,

$$\varphi(G) = \sup\ \{\varphi(B) : B \subseteq E - F,\ B \text{ closed}\},$$

one sees that $(F, G) \in \mathcal{E}$ or $(G, F) \in \mathcal{E}$. Let $\mathcal{E}_1$ be the class of all E_1 such
that $(E_1, E_2) \in \mathcal{E}$ for some E_2, and set $\alpha = \inf\{\varphi(E_1)/\varphi(E) : E_1 \in \mathcal{E}_1\}$.

If $\alpha < 2/3$ the truth of the Proposition is evident. Suppose then
that $\alpha \geqslant 2/3$. There are two possibilities:

Case a. There exist $(E_1', E_2') \in \mathcal{E}$ and $(E_1'', E_2'') \in \mathcal{E}$ such that
$\varphi(E_1' \cap E_1'') < \alpha \varphi(E)$. Note that $E - (E_1' \cap E_1'')$ contains the closed
set E_2', and hence it contains a largest closed set G. Evidently
$\varphi(G) \leqslant 2\varphi(E)(1-\alpha) \leqslant (2/3)\varphi(E)$. It follows that $(G, E_1' \cap E_2'') \in \mathcal{E}$

and $\varphi(G) = (2/3)\varphi(E) = a\varphi(E)$. So in this case the conclusion of the Proposition holds.

Case b. For all $E_1' \in \mathfrak{S}_1$, $E_1'' \in \mathfrak{S}_1$ one has $\varphi(E_1' \cap E_1'') \geqslant a\varphi(E)$. By definition of a, there exist $E_1^i \in \mathfrak{S}_1$ with $\varphi(E_1^i) \leqslant (a + in^{-1})\varphi(E)$, $i = 1, 2, \ldots$. Let $E_1 = \bigcap_{i=1}^{\infty} E_1^i$. Using the assumption of Case b, one sees that $\varphi(E_1) = a\varphi(E)$. It is easy to see that there exists an E_2 such that $(E_1, E_2) \in \mathfrak{S}$. Since E_1 is not indecomposable it contains disjoint closed subsets E_{11} and E_{12} such that $E_1 - (E_{11} \cup E_{12})$ contains no closed subset, $\varphi(E_{12}) \leqslant (a/2)\varphi(E)$, $\varphi(E_{12}) < a\varphi(E)$. One easily sees that $E - (E_{11} \cup E_{12} \cup E_2)$ contains no closed subset. Hence $(E_{12} \cup E_2, E_{11}) \in \mathfrak{S}$ or $(E_{11}, E_{12} \cup E_2) \in \mathfrak{S}$. However $\varphi(E_{11}) < \varphi(E_1) = a\varphi(E)$, so the second alternative is ruled out and $(E_{12} \cup E_2, E_{11}) \in \mathfrak{S}$. Now $\varphi(E_{12} \cup E_{11}) \leqslant [(a/2) + (1-a)]\varphi(E) \leqslant (2/3)\varphi(E)$ since $a \geqslant (2/3)$; but this implies that $a = 2/3$ and all the inequalities can be replaced by equalities, and again the desired conclusion is seen to hold.

PROPOSITION 9.6. *Consider a φ-non-singular chain on $(S,\mathfrak{B})$. Let E be closed and suppose E contains no indecomposable set. Then E is not properly essential.*

Proof. It may be assumed that φ is finite, since in any case there exists an equivalent finite measure. By the previous proposition E contains two closed disjoint subsets E_1 and E_2 such that $E - (E_1 \cup E_2)$ contains no closed subset and $\varphi(E_j) \leqslant (2/3)\varphi(E)$, $j = 1, 2$. In the same way for $i = 1$ and 2, E_i gives rise to two closed subsets E_{i1}, E_{i2} of E_i such that $E_i - (E_{i1} \cup E_{i2})$ contains no closed subsets, $\varphi(E_{ij}) \leqslant (2/3)\varphi(E_i)$, $j = 1, 2$. Proceding in this way one obtains $E_{i_1 i_2, \ldots, i_n}$, $i_1 \in \{1,2\}$, $i_2 \in \{1,2\}$, $\ldots$, $i_n \in \{1,2\}$, $n = 1, 2, \ldots$. Let $F_n = \bigcup_{i_1, i_2, \ldots, i_n} E_{i_1 i_2 \cdots i_n}$. Let $F = \bigcap_{n=1}^{\infty} F_n$.

If F is not void, choose $x \in F$. There exist then a sequence of indexes i_1, i_2, ..., such that $x \in E_{i_1 i_2 \cdots i_n}$ for each n. By construction $\varphi(E_{i_1 i_2 \cdots i_n}) \leq (2/3)^n \varphi(E)$ so that $\bigcap_{n=1}^{\infty} E_{i_1 i_2 \cdots i_n}$ is a closed set of φ measure 0. This contradicts the φ-non-singularity and shows $F = \phi$. Therefore

$$E = (E - F_1) \cup (F_1 - F_2) \cup \ldots.$$

Since each summand on the right is a set which is not properly essential, it follows that E is improperly essential.

THEOREM 9.1. *Consider a φ-non-singular Markov chain on $(S, \mathcal{B})$. Then $S = \dot{\cup} L_i \dot{\cup} I$, where the L_i are the at most denumerable many maximal indecomposable sets and I is not properly essential.*

Proof. Again it may be assumed that φ is finite. Let $L = \bigcup_i L_i$ where the L_i run through the maximal indecomposable sets; (see Proposition 9.2, Proposition 9.3). Since L is closed, $L^0 \cap L = \phi$. Note that $S - (L \cup L^0)$ can contain no closed set, so it is not properly essential by Proposition 9.4. If L^0 is void there is nothing left to prove. So suppose $L^0 \neq \phi$. It follows from Proposition 9.2 that L^0 contains no indecomposable set. So Proposition 9.6 applies to show that L^0 is not properly essential, and the theorem is proved.

An immediate consequence is the

COROLLARY. *For a φ-non-singular Markov chain on $(S, \mathcal{B})$, $S = \bigcup_i L_i' \cup J$ where L_i' ranges over all properly essential maximal indecomposable sets and J is not properly essential.*

Putting together the Corollary to Theorem 9.1 with Theorem 8.2 gives

THEOREM 9.2. *Consider a φ-non-singular chain on $(S, \mathcal{B})$. If $\mathcal{B}$ is separable and S is properly essential*

$$S = \cup_i H_i \cup I$$

where the union is over a non-void index set which is at most denumerable and each H_i is φ-recurrent.

CHAPTER 2

Ratios and differences of sums of transition probabilities

Throughout this chapter (X_i) is a Markov chain on $(S, \mathcal{B})$ which is φ-recurrent. Let π denote some fixed non-trivial invariant σ-finite measure. Then the chain is π-recurrent.

For $A \in \mathcal{B}$, $B \in \mathcal{B}$ and μ, ν two probability measures on $(S, \mathcal{B})$, let

(1) $$R^n(\mu, A; \nu, B) = \frac{\sum_{k=0}^{n} \int P^k(x, A) \mu(dx)}{\sum_{k=0}^{n} \int P^k(y, B) \nu(dy)}.$$

In case $\mu = \epsilon_x$ or $\nu = \epsilon_y$ we may write x, or y respectively in the corresponding argument place; e.g. $R^n(x, A; y, B)$ has the same significance as $R^n(\epsilon_x, A; \epsilon_y, B)$. The right side of (1) may have a vanishing denominator. However, if $\pi(B) > 0$, the denominator will be positive for all large enough n, and this is enough for our purposes.

If π is a probability measure, divide both numerator and denominator of the right side of (1) by n and observe that Theorem 7.1 of Chapter 1 implies that as n tends to infinity one obtains a limit equal to $\pi(A)$ in the numerator and $\pi(B)$ in the denominator. Of principle interest to us will be the case when π is not finite: in that case Theorem 7.3 of Chapter 1 shows that normalizing numerator and denominator by the factor n^{-1} and passing to the limit will give a quotient of the form $0/0$ whenever $\pi(B) < \infty$, $\pi(A) < \infty$. Nevertheless, as will be seen, under suitable conditions the right side of (1) can be shown to converge to $\pi(A)/\pi(B)$ as $n \to \infty$.

PROPOSITION 1. *Let D be a D-set, μ, ν probability measures on $(S, \mathcal{B})$. Then $\lim_{n \to \infty} R^n(\mu, D; \nu, D) = 1$.*

Proof. Let π_D be the invariant probability measure for the process on D. Let $H_n(x) = \sum_{k=0}^{n} P^k(x, D)$, and for any probability measure ν on $(S, \mathcal{B})$ put $H_n(\nu) = \int H_n(x) \nu(dx)$. Since $R^n(\mu, D; \nu, D) = R^n(\mu, D; \pi_D, D)/R^n(\nu, D; \pi_D, D)$ for big enough n, it evidently suffices to prove that for each probability measure μ $R^n(\mu, D; \pi_D, D)$ converges to 1. Write H_n for $H_n(\pi_D)$. Then what is wanted is $H_n(\mu)/H_n$ approaches 1.

Let $\delta > 0$. Set $B_{n, \delta} = \{x \in D : H_n(x) \leqslant (1 + \delta) H_n\}$. Then

$$H_n \geqslant \int_{D - B_{n, \delta}} H_n(x) \pi_D(dx) \geqslant (1 + \delta) H_n \pi_D(D - B_{n, \delta})$$
$$= (1 + \delta) H_n (1 - \pi_D(B_{n, \delta})),$$

and provided only that n is big enough so that H_n is positive, one obtains

(2) $$\pi_D(B_{n, \delta}) \geqslant \frac{\delta}{1 + \delta},$$

the important thing being that the lower bound is independent of n.

If the Markov chain has initial probability distribution ν, $H_n(\nu)$ is simply the expected number of visits to D up to time n, and this is the sum of the expected number of visits paid D up to time n prior to entering $B_{n, \delta}$, and the expected number of visits paid D by time n after entering $B_{n, \delta}$. The last-mentioned expectation is bounded by $(1 + \delta)n$. Hence

(3) $$H_n(\nu) \leqslant \sum_{k=0}^{n} \int_{B_{n,\delta}} P^k(x,D)\nu(dx) + (1+\delta)H_n$$
$$\leqslant \int \sum_{k=0}^{\infty} {}_{B_{n,\delta}} P^k(x,D)\nu(x) + (1+\delta)H_n.$$

The first term of the last member is the expected value, for the chain with initial probability distribution ν, of the number of visits to D prior to entering $B_{n,\delta}$, and it can be seen that this quantity is bounded by a positive constant K_δ not depending on ν or n. To see this note that the process on D is uniformly π_D-recurrent, and apply (iii) of Theorem 7.1 of Chapter 1. Choose n_δ so that $K_\delta \leqslant \delta H_{n_\delta}$. Then (3) results in

(4) $$H_n(\nu) \leqslant (1+2\delta)H_n, \qquad n \geqslant n_\delta,$$

for all initial probability distributions ν. So one obtains

(5) $$\limsup_{n \to \infty} H_n(\mu)/H_n \leqslant 1.$$

Let $C_{n,\delta} = \{x \in D : H_n(x) \geqslant (1-\delta)H_n\}$. Use (4) with $\nu = \epsilon_x$ to obtain that for $n \geqslant n_\delta$

$$H_n = \int_{C_{n,\delta}} H_n(x)\pi_D(dx) + \int_{D-C_{n,\delta}} H_n(x)\pi_D(dx)$$
$$\leqslant (1+2\delta)\pi_D(C_{n,\delta}) + (1-\delta)H_n\pi_D(D-C_{n,\delta}),$$

and so

$$1 \leqslant (1+2\delta)\pi_D(C_{n,\delta}) + (1-\delta)(1-\pi_D(C_{n,\delta})) = (1-\delta) + 3\delta\,\pi_D(C_{n,\delta}),$$

and thus

(6) $$\pi(C_{n,\delta}) \geqslant 1/3, \qquad n > n_\delta.$$

Observe that

$$H_{n+m}(\mu) \geq \sum_{\nu=1}^{m} \int_{y \in C_{n,\delta}} [\int_{x \in S} C_{n,\delta} P^\nu(x, dy) \mu(dx) \cdot \sum_{s=0}^{n+m-\nu} P^s(y, D)]$$

$$\geq \inf_{y \in C_{n,\delta}} H_n(y) \cdot \sum_{\nu=1}^{m} \int_{C_{n,\delta}} P^\nu(x, C_{n,\delta}) \mu(dx)$$

for every n and m, and therefore

(7) $$H_{n+m}(\mu) \geq (1-\delta) H_n \cdot \sum_{\nu=1}^{m} \int_{C_{n,\delta}} P^\nu(x, C_{n,\delta}) \mu(dx).$$

Again using the fact that the process on D is uniformly π_D-recurrent, it follows from (6) that there exists an m_0, depending on δ but not on n or on the initial distribution of the chain, such that the probability of visiting $C_{n,\delta}$ by the time D has been visited m_0 times exceeds $(1-\delta)$ provided $n \geq n_\delta$. Since the chain is π-recurrent there exists an m_1, which may depend on the initial probability measure μ, such that P_μ [entering D at least m_0 times by time m_1] $> 1-\delta$. Therefore P_μ [visiting $C_{n,\delta}$ by time m_1] $\geq P_\mu$ [entering D at least m_0 times by time m_1 and visiting $C_{n,\delta}$ by the time D has been visited m_0 times] $\geq 1-2\delta$. Therefore it follows from (7) that for $n \geq n_\delta$,

(8) $$H_n(\mu) + m_1 \geq H_{n+m_1}(\mu) \geq (1-\delta)(1-2\delta) H_n.$$

It follows that

(9) $$\liminf_{n \to \infty} H_n(\mu)/H_n \geq (1-\delta)(1-2\delta),$$

and since δ is arbitrary (9) and (5) imply the theorem.

PROPOSITION 2. *Let* $A, B \in \mathcal{B}$, $0 < \pi(B) < \infty$. *Then* $\lim_{n \to \infty} R^n(x, A; x, B) = \pi(A)/\pi(B)$ *as n approaches infinity for π-a.e. x.*

Proof. Suppose first that $\pi(A) < \infty$. Then the result is an immediate consequence of the Chacon-Ornstein theorem. (See Neveu [1964]). For let $L_1 = L_1(S, \mathcal{B}, \pi)$, and for $f \in L_1$ let $Pf(x) = \int P(x, dy) f(y)$. Then P is a positive contraction on L_1, so that the Chacon-Ornstein Theorem is applicable. Applying the theorem to the L_1 functions $I_A(\cdot)$ and $I_B(\cdot)$ one obtains the π-a.e. convergence of $R^n(x, A; x, B)$. Evidently for $f \in L_1$ and $f \geq 0$, $\sum_{n=0}^{\infty} P^n f(x)$ equals 0 or ∞, that is P is conservative; and any closed set $E \in \mathcal{B}$ satisfies $\pi(S - E) = 0$, so that P is ergodic. Therefore the Chacon-Ornstein Theorem also tells us that the limit of $R^n(x, A; x, B)$ equals $\pi(A)/\pi(B)$ for π-a.e. x. Finally the case $\pi(A) = \infty$ follows by a limiting argument.

Remark. Proposition 2 and Theorem 1 below are a.e. results, in distinction to nearly all the other results in these notes.

THEOREM 1. *For $A, B \in \mathcal{B}$, $0 < \pi(B) < \infty$, there exists a π-null set $N_0(A, B)$ such that*

$$\lim_{n \to \infty} R^n(x, A; y, B) = \frac{\pi(A)}{\pi(B)}, \qquad x \in S - N_0(A, B), \qquad y \in S - N_0(A, B).$$

Proof. Let D be a D-set. Write

(10) $R^n(x, A; y, B) = R^n(x, A; x, D) R^n(x, D; y, D) R^n(y, D; y, B),$

the relation holding for all big enough n. By Proposition 2, as n tends to infinity, on the right side of (10) the first factor tends to $\pi(A)/\pi(D)$ provided x does not belong to a π-null set $N(A, D)$; and the last factor converges to $\pi(D)/\pi(B)$ provided y does not belong to the π-null set $N(D, B)$. By Proposition 1 the middle factor converges to 1. Hence the theorem follows with $N_0(A, B) = N(A, D) \cup N(D, B)$.

It should be noted that for any $A \in \mathcal{B}$, if $A \subseteq D$, for some D-set D, then either $\pi(A) = 0$ or A is also a D-set.

THEOREM 2. *Let B be a D-set, and let $A \in \mathcal{B}$ be included in a D-set. Then for every pair of initial probability measures μ, ν, $\lim_{n \to \infty} R^n(\mu,A;\nu,B) = \pi(A)/\pi(B)$.*

Proof. Suppose $A \subseteq D$, where D is a D-set. If A has positive π-measure it is a D-set; otherwise $D - A$ is a D-set. Thus it will suffice to prove the theorem under the assumption that A is a D-set. Choose $z \in S$ outside the exceptional null set corresponding to (A, B) by Proposition 2. For all big enough n

$$R^n(\mu,A;\nu,B) = R^n(\mu,A;z,A)R^n(z,A;z,B)R^n(z,B;\nu,B),$$

and the truth of the theorem becomes apparent on applying Proposition 1 and Proposition 2 to the right side of the last relation.

Consider Theorem 1 in the denumerable case. Let $N = \{x : \pi(\{x\}) = 0\}$. Evidently every π-null set is a subset of N. In particular the convergence assertion of Theorem 1 holds true for all x and y outside of N. The question arises whether Theorem 1 can be improved to assert the existence of a fixed null set N, independent of A and B, so that the convergence takes place for all $x \in S - N$, $y \in S - N$. The following example will show that no such improvement is possible.

Example. The state space S consists of the union of the following linear sets: $R = \{x : -\infty < x < \infty\}$; $A_n = \{(x,n) : 0 \leqslant x \leqslant 1\}$, $n = 1, 2, \ldots$ and $\mathcal{B}$ is the σ-field generated by the ordinary Borel sets on the indicated segments. First we describe a transition probability P_0 for $x \in R$ only: if $x \in [n, n+1)$, $P_0(x, \cdot)$ is uniformly distributed on $[n-1, n) \cup [n+1, n+2)$. The transition probability

SUMS OF TRANSITION PROBABILITIES 55

$P(\cdot,\cdot)$ on S is defined as follows:

$$P(x,\cdot) = P_0(x,\cdot) \text{ for } x \in R - [0,1)$$
$$P(x,\cdot) = \frac{1}{2} P_0(x,\cdot) + \frac{1}{2} \epsilon_{(x,1)}, \text{ for } x \in [0,1)$$
$$P((x,n),\cdot) = a_n \epsilon_{(x,n+1)} + (1-a_n)P_0(x,\cdot), \text{ for } x \in [0,1), \quad n=1,2,\ldots,$$

where the constants a_n will be specified further. Let $\beta_1 = 1/2$ and for $n > 1$, let $\beta_n = 1/2(a_1 a_2 \ldots a_{n-1})$. The a_n are to satisfy (i) $0 < a_n < 1$, (ii) $\lim_{n \to \infty} \beta_n = 0$, (iii) $\beta_n > \sup_{0 \leq x < 1} P_0^n(x,[0,1))$. It should be noted that such a choice of a_n is possible. Note also that $P(\cdot,\cdot)$ determines a chain which is φ-recurrent, where φ is Lebesgue measure on R, $\varphi(S-R) = 0$. Hence there exists an invariant σ-finite measure π. Let $B_x = \{(x,n) : 1 \leq n < \infty\} \cup \{x\}$, where $x \in [0,1)$. Then for $n \geq 1$,

$$\frac{\sum_{k=0}^{n} P^k(x, B_x)}{\sum_{1}^{n} P_0^k(x,[0,1))} = \frac{1 + \sum_{k=1}^{n} \beta_k}{1 + \sum_{k=1}^{n} P_0^k(x,[0,1))} \geq 1.$$

Now

$$\sum_{k=0}^{n} P_0^k(x,[0,1)) \geq \sum_{k=0}^{n} P^k(x,[0,1))$$

because $P_0(x,\cdot)$ is exactly the transition probability function for the process on R: thus if the initial distribution is ϵ_x the right side represents the expected number of visits the chain pays to $[0,1)$ in n steps, while the left side represents the expected number of visits paid to $[0,1)$ during the first n visits to R, and the asserted inequality is evident. Thus

$$\liminf_{n \to \infty} \frac{\sum_{k=0}^{n} P^k(x, B_x)}{\sum_{k=0}^{n} P^k(x, [0,1))} \geqslant 1,$$

and since evidently $\pi[0,1) > 0$, $\pi(B_x) = 0$ it follows that each $x \in [0,1)$ belongs to the exceptional set $N_0(B_x, [0,1))$ of Theorem 1.

However, a positive result along the lines sought also exists.

THEOREM 3. *For $T \in \mathcal{B}$ with $0 < \pi(T) < \infty$ there exists $N(T) \in \mathcal{B}$ with $\pi(N(T)) = 0$ such that for all $A \in \mathcal{B}$ and $B \in \mathcal{B}$ with $0 < \pi(B)$, $A \subset T$ and $B \subset T$ one has*

$$\lim_{n \to \infty} R^n(x, A; y, B) = \frac{\pi(A)}{\pi(B)}, \qquad x \in S - N(T), \qquad y \in S - N(T).$$

Proof. According to Theorem 6.2 of Chapter 1, $S = \bigcup_{m=1}^{\infty} D_m$, where (D_m) is an increasing sequence of D-sets. It will suffice to show the existence of a π-null set $N(T)$ such that for every $A \in \mathcal{B}$ such that $A \subset T$ and every $x \in S - N(T)$, $y \in S - N(T)$, $R^n(x, A; y, D_1)$ converges to $\pi(A)/\pi(D_1)$. Write, for big enough n,

(10) $R^n(x, A; y, D_1) = R^n(x, A \cap D_m; y, D_1) + R^n(x, A - D_m; y, D_1).$

According to Theorem 2, as n tends to infinity the first term on the right of (10) tends to $\pi(A \cap D_m)/\pi(D_1)$, and this in turn approaches $\pi(A)/\pi(D_1)$ as m goes to infinity. The last term in (10) is bounded by $R^n(x, T - D_m; y, D_1)$. It follows from Theorem 1 that there exists a π-null set N_T such that for x and y outside N_T

$\lim_{n \to \infty} R^n(x, T - D_m; y, D_1) = \pi(T - D_m)/\pi(D_1)$ for $m = 1, 2, \ldots$. Since $\pi(T - D_m)$ approaches 0 as m tends to infinity it now follows that as n approaches infinity the left side of (10) converges to $\pi(A)/\pi(D_1)$, as had to be shown.

, When can one assert that the quantity:

$$(11) \quad D^n(\mu, A; \nu, B) = \pi(B) \sum_{k=1}^{n} \int P^k(x, A) \mu(dx) - \pi(A) \sum_{k=1}^{n} \int P^k(y, B) \nu(dy),$$

remains bounded in absolute value as n tends to infinity, or perhaps even that convergence takes place?

PROPOSITION 3. *Suppose (X_n) is uniformly π-recurrent and aperiodic. Let f be a bounded measurable function satisfying*

$$\int f(x) \pi(dx) = 0.$$

Then as n tends to infinity

$$\sum_{k=1}^{n} P^k f(x) \text{ converges uniformly in } x.$$

Proof. The result is a consequence of (iii) of Theorem 7.1 in Chapter 1.

Returning now to the general situation where only π-recurrence is assumed, one has the following theorem.

THEOREM 4. *Let D be a D-set, f a bounded measurable function vanishing on $S - D$ and satisfying:*

$$(12) \qquad \int f(x) \pi(dx) = 0.$$

Then there exists a constant M such that

$$\left| \sum_{j=1}^{n} P^j f(x) \right| < M, \qquad n = 0, 1, \ldots, \; x \in S,$$

Proof. Suppose first that the process on D is aperiodic. Let

$$(13) \qquad K_n(x, A) = \sum_{k=1}^{n} P_D^k(x, A),$$

and note that it follows from Proposition 3 that

$$Kf(x) = \lim_{n \to \infty} \int K_n(x, dy) f(y)$$

is well defined. For $B \subseteq D$ evidently

$$P_D^k(x, B) = \int_{S-D} P(x, dy) P_D^k(y, B) + \int_D P(x, dy) P_D^{k-1}(y, B),$$

where $P_D^0(y, B) = I_B(y)$, and this results in

$$P[P_D^k f] = P_D^k f + \int_D P(x, dy) [P_D^k f(y) - P_D^{k-1} f(y)].$$

Summing k from 1 to n, and transposing, one obtains

(14) $$(I-P)K_n f = Pf - \int_D P(\cdot, dy) P_D^n f(y),$$

where, of course, I is the identity, that is,

$$I(x, A) = I_A(x), \quad x \in S, \quad A \in \mathcal{B}.$$

As n tends to infinity the last term in (14) goes to zero, which is a direct consequence of (12), obtained from (iii) of Theorem 7.1, Chapter 1. So one obtains

(15) $$(I-P)Kf = Pf,$$

and therefore

(16) $$\sum_{j=1}^n P^j f = Kf - P^n Kf.$$

Since Kf is evidently bounded, the desired conclusion follows from (16).

It still remains to treat the case in which the process on D has period $d > 1$. Let $C_1, C_2, \ldots, C_d$ be a cycle for the process on D as guaranteed by Theorem 3.1 of Chapter 1. Let $F = D - (C_1 \cup C_2 \ldots \cup C_d)$. Then $\pi(F) = 0$. Evidently

$\sum_{n=1}^{\infty} P^n(x,F) = \sum_{n=1}^{\infty} P_D^n(x,F)$ and (iii) of Theorem 7.1 of Chapter 1 shows that the last sum is bounded uniformly in x. So the conclusion of the theorem holds if f vanishes on $S - F$. Since $f = f \cdot I_F + f \cdot I_{S-F}$ it will be enough to consider functions f vanishing on F. Let

$$a_i = \int_{C_i} f(x) \, \pi \, (dx)/\pi(C_i),$$

and note that (12) implies $\sum_{i=1}^{d} a_i = 0$. Since the process on C_i is aperiodic the theorem is already established for the functions

$$f_i = I_{C_i}(f_i - a_i).$$

Now $f = \sum_{i=1}^{d} f_i + \sum_{i=1}^{d} a_i I_{C_i}$, so it remains only to treat functions of the form

$$f = \sum_{i=1}^{d} a_i I_{C_i}, \qquad \sum_{i=1}^{d} a_i = 0.$$

Since the process on D moves through the C_i in order,

$$\sum_{k=1}^{n} f(X_k) \text{ must equal } \sum_{i=q}^{m} a_i$$

for some m and q with $0 \leqslant q \leqslant m \leqslant d$, and the conclusion of the theorem is evidently true.

COROLLARY. *Let D be a D-set, $A \subseteq D$, $B \subseteq D$, $A \in \mathcal{B}$, $B \in \mathcal{B}$. Then there exists an M such that $|D^n(\mu, A; \mu, B)| \leqslant M$ for all $n \geqslant 0$ and all initial probability measures μ.*

Proof. Let $f = \pi(B)I_A - \pi(A)I_B$. Since

$$|D^n(\mu, A; \mu, B)| \leqslant \sup_{x \in S} \left| \sum_{j=1}^{n} P^j f(x) \right|$$

the Corollary follows from Theorem 4.

The conclusion of the corollary is of course stronger than the assertion that $R^n(\mu,A;\mu,B)$ tends to $\pi(A)/\pi(B)$ as n approaches infinity. On the other hand Theorem 2 is stronger in that it allows distinct initial probabilities μ and ν. In general it does not seem possible to drop the restriction $\mu = \nu$ in the last corollary. However, the next theorem is a result in that direction.

THEOREM 5. *Let D be a D-set and put*

$$H_n(x) = \sum_{k=1}^{n} P^k(x, D), \quad H_n^* = \sup_x H_n(x),$$
$$C_{n,N} = \{x \in D : H_n(x) \leq H_n^* - N\}.$$

Then as N approaches infinity $\pi(C_{n,N})$ tends to zero uniformly in n.

Proof. Let T_1, T_2, ... be the times of first, second, ... visit of (X_n) to D. Suppose $\pi(C_{n,N}) \geq \delta > 0$. It follows from (iii) of Theorem 7.1 of Chapter 1 that there exists a constant K_δ, depending only on δ, such that

$$H_n(x) \leq \sum_{k=1}^{\infty} P_x[X_{T_k} \in C_{n,N}, X_{T_j} \notin C_{n,N}, j < k] (H_n^* - N + k)$$
$$\leq H_n^* - N + K_\delta.$$

Taking the supremum over x it follows that $N \leq K_\delta$ and the theorem is proved.

COROLLARY. *Let $A \in \mathcal{B}$, $B \in \mathcal{B}$, $A \subseteq E$, $B \subseteq E$, where E is a D-set. Suppose that $x \in S$ and $y \in S$ satisfy $\pi(\{x\}) > 0$, $\pi(\{y\}) > 0$. Then $|D^n(\epsilon_x, A, \epsilon_y, B)|$ is bounded uniformly in n.*

Proof. Observe that $D = E \cup \{x, y\}$ is a D-set. Theorem 4 allows one to reduce the assertion to the case $A = B = D$. Now Theorem 5 and its notation will be used. Since $\pi(C_{n,N})$ converges to zero as N approaches infinity, it follows that for N sufficiently large $x \notin C_{n,N}$ and $y \notin C_{n,N}$. Then

$$D^n(\epsilon_x, D, \epsilon_y, D) \leqslant \pi(D)(|H_n(x) - H_n^*| + |H_n^* - H(y)|)$$
$$\leqslant \pi(D)\, 2\, N.$$

The corollary is mainly of interest for denumerable state space.

Finally, we consider the convergence of $D^n(\mu, A; \mu, B)$ as n tends to infinity: Using the notation introduced in (13), note that if the process on D is aperiodic

$$\int P^n(x, dy) K_N f(y) = \int_{S-D} P^n(x, dy) \left[\int P_D(y, dz)(f(z) + K_{n-1} f(z)) \right]$$
$$+ \int_D P^n(x, dy) K_n f(y),$$

and letting N tend to infinity one obtains:

(17) $\quad \int P^n(x, dy) K f(y) = \int_{S-D} P^n(x, dy) \left[\int P_D(y, dz)(f(z) + K f(z)) \right]$
$$+ \int_D P^n(x, dy)\, K f(y).$$

If one sets:

$$H_D(x, B) = I_{S-D}(x) P_D(x, B\ D) + I_{D \cap B}(x),$$

then $H_D(x,B)$ represents the probability that the chain finds itself in B when it first enters D, given that it starts at x; that is

$$H_D(x,B) = P_x[\bigcup_{n=0}^{\infty} [X_n \in B \cap D, \ X_i \notin D, \ i = n\text{-}1, \ n\text{-}2, \ldots, 0]].$$

Now (17) can also be written as

(18) $\quad \int P^n(x,dy) \, K f(y) = \int P^n(x,dy) \, [H_D K f(y)$

$$+ H_D \, f(y)] - \int_D P^n(x,dy) \, f(y).$$

THEOREM 6. *Let D be a D-set and f a bounded measurable function vanishing on $S-D$ and satisfying (12). Then*

(19) $\qquad \sum_{j=1}^{n} P^j f$ *converges as* $n \to \infty$

provided

(20) $\quad \int P^n(x,dy) \, H_D(y,B)$ *converges as* $n \to \infty$ *for every* $B \in \mathcal{B}$.

Proof. Suppose first that the process on D is aperiodic. Then (16) applies to show that (19) is equivalent to

(21) $\qquad\qquad P^n \, K f$ converges as $n \to \infty$.

Suppose now that (20) holds. Then the first term on the right in (18) must converge. In fact the second term must also converge. When $\pi(S) = \infty$ this follows from Theorem 7.3 of Chapter 1. When $\pi(S) = 1$ use (i) of Theorem 7.1 of Chapter 1, observing that in this case (20) implies aperiodicity of the underlying Markov chain. So (21) is established. If the process on D has period d greater than 1 one makes use of the cyclic decomposition much as in the proof of Theorem 4.

Note that if (X_n) is aperiodic and $\pi(S) = 1$, (20) is always true, as one sees easily from (i) of Theorem 7.1 of Chapter 1. The more interesting case of course is $\pi(S) = \infty$. Then (19) is not necessarily true. There are, however, important examples where (20) does hold. Particular mention must be made of random walks (see Example 1.3 of Chapter 3), for which (20) or suitable analogues have been established.

CHAPTER 3

Individual ratio limit theorems

1. Introduction and first results. It is natural to attempt to improve upon the results of the previous chapter by studying the limiting behaviour of the *individual ratios* $P^{n+m}(x,A)/P^n(x,B)$, where n tends to infinity, m is a fixed integer, $x, y \in S$ and $A, B \in \mathcal{B}$. For aperiodic φ-recurrent chains with invariant measure π it is reasonable to expect the limit to exist and equal $\pi(A)/\pi(B)$. It will be seen that the problem is not yet fully understood, and that the results here are much more fragmentary than those of the previous two chapters. Indeed much of our discussion will proceed in the much more restrictive framework of the following hypothesis:

(D) The state space $(S, \mathcal{B})$ of the Markov chain consists of an at most denumerable set S, and the Borel field $\mathcal{B}$ is generated by the sets $\{s\}$, $s \in S$.

Even in the denumerable case many unsolved problems remain. Frequently we require the hypothesis:

(R) The chain is φ-recurrent for some φ.

As we know from Chapter 1, (R) implies that there exists a σ-finite invariant measure π, unique up to constant multiple, and the chain will be π-recurrent. The notation π will be reserved for such a measure.

When (D) is assumed, we make the notational convention that a function argument for a set of $\mathcal{B}$ can be occupied by the name of a state of S with the understanding that the name of the state designates the corresponding singleton, e.g., $\pi(i)$ for $\pi(\{i\})$, $_kP(i,j)$ for $_{\{k\}}P(i,\{j\})$.

If both (D) and (R) hold $\{i \in S : \pi(i) > 0\}$ is a closed set. The process on this set satisfies the condition:

(DR) $\qquad$ Both (D) and (R) hold and $\pi(i) > 0$ for all i.

Frequently it is convenient to assume (DR). Finally, we often will use:

(A) $\qquad\qquad\qquad$ The process is aperiodic.

This condition will be used when (R) is assumed; for the definition of aperiodic see Chapter 1, end of section 3.

When dealing with the denumerable case the notation now to be introduced will be used throughout this chapter. Let 0 be a fixed state of S and write:

(1.1) $$u_n = P^n(0,0), \quad n = 0, 1, \ldots.$$

Recall that this makes $u_0 = 1$. Set

(1.2) $$f_k = {_0P^k}(0,0), \qquad k = 1, 2, \ldots,$$

so that f_k is the probability of returning to zero for the first time in k steps. Evidently

(1.3) $$0 \leq f_k \leq 1,$$

(1.4) $$\sum_{k=1}^{\infty} f_k \leq 1.$$

If the Markov chain satisfies (R) then

$$(1.5) \qquad \sum_{k=1}^{\infty} f_k = 1,$$

and if furthermore (A) is satisfied, then

(1.6)   the greatest common divisor of $\{k : f_k > 0\}$ is one.

The f_k and the u_n are connected by the *renewal equation*

$$(1.7) \qquad u_n = \sum_{k=1}^{n} f_k u_{n-k}, \qquad n = 1, 2, \ldots, \qquad u_0 = 1.$$

Starting with any set of f_k satisfying (1.3) and (1.4) one can always find a Markov chain satisfying (D) so that (1.2) and (1.1) hold; and if furthermore (1.5) and (1.6) hold, the chain can be chosen to satisfy (DR) and (A). This is very easy to verify; for details see Example 1.1 below.

When (DR) holds, the convergence of the individual ratios is intimately connected with the behavior of u_{n+1}/u_n; the connection will be made more precise below.

Some well known consequences of the renewal equation are summarized in the next proposition.

PROPOSITION 1.1. *Let* (f_k), $k = 1, 2, \ldots$ *satisfy (1.3) and (1.4), and let the u_n be defined by (1.7). Then*

$$(1.8) \qquad 0 \leqslant u_n \leqslant 1, \qquad n = 0, 1, \ldots.$$

$$(1.9) \qquad u_{n+m} \geqslant u_n \cdot u_m, \qquad n = 0, 1, \ldots, \qquad m = 0, 1, \ldots.$$

$$(1.10) \qquad \sum_{n=0}^{\infty} u_n = \infty \text{ if and only if } \sum_{k=1}^{\infty} f_k = 1.$$

If (1.6) holds

(1.11) $\quad u_n > 0$ *for all sufficiently big n,*

and if also (1.5) holds

(1.12) $$\lim_{n \to \infty} u_n = (\sum_{k=1}^{\infty} k\, f_k)^{-1},$$

where the right side of (1.12) is 0 when the sum diverges.

Condition (1.6) implies

(1.13) $$0 < \lim_{n \to \infty} (u_n)^{1/n} = \rho \leqslant 1$$

exists, and if also (1.5) holds $\rho = 1$.

Proof. As already mentioned we may suppose the representation (1.1) and (1.2) to hold. Then (1.8) and (1.9) are obvious. The assertion (1.10) is most quickly proved by noting that if

$$U(s) = \sum_{n=0}^{\infty} u_n\, s^n, \quad F(s) = \sum_{k=1}^{\infty} f_k\, s^k, \qquad 0 \leqslant s < 1,$$

then $U(s) = (1 - F(s))^{-1}$ and letting s converge up to 1.

That (1.6) implies (1.11) can be considered a special case of Proposition 2.1 of Chapter 1 (with $\{0\}$ as the C-set, ϵ_0 as φ).

Assume (1.6) and (1.5); then the chain may be assumed to satisfy (DR) and (A). Let π be the invariant measure normalized so that $\pi(0) = 1$. According to the construction of π given in Theorem 7.2 of Chapter 1 (see especially (7.2)) $\pi(S)$ is the expected return time to 0, that is $\pi(S) = \sum_{k=1}^{\infty} k\, f_k$. If this sum is finite $\pi'(i) = \pi(i)/\pi(S)$ is an invariant probability measure and then (1.12) is an instance of Theorem 7.1 of Chapter 1. When $\sum_{k=1}^{\infty} k\, f_k = \pi(S) = \infty$, (1.12) follows from Theorem 7.3 of Chapter 1.

To see that (1.6) implies (1.13) suppose first that $f_1 > 0$. Then $u_n > 0$ for all n. Let $a_n = -\log u_n$. It follows from (1.8) and (1.9) that

$$0 \leqslant a_{n+m} \leqslant a_n + a_m < \infty$$

and for such a sequence one proves easily that

$$\lim_{n \to \infty} \frac{a_n}{n} = \liminf_{n \to \infty} \frac{a_n}{n} < \infty.$$

In case $f_1 = 0$ we use a device which will be useful frequently below: suppose $f_{N_1} > 0$, $f_i = 0$, $1 \leqslant i < N_1$. Looking at the Markov chain only at times $0, N_1, 2N_1, \ldots$, one obtains another chain with transition probabilities $\tilde{P}$ such that $\tilde{P}^k(i,j) = P^{kN_1}(i,j)$. In particular $\tilde{u}_n = u_{nN_1}$. The first return times $(\tilde{f}_k)$ satisfy $\sum_{k=1}^{\infty} \tilde{f}_k = 1$ if and only if $\sum_{k=1}^{\infty} f_k = 1$. But now $\tilde{f}_1 = f_{N_1} > 0$, so the argument above applies to give $(\tilde{u}_n)^{1/n} = (u_{nN_1})^{1/n}$ converges to a positive limit; hence $(u_{nN_1})^{1/(nN_1)}$ converges to a positive limit as $u \to \infty$. Using (1.6) and its consequence (1.11) together with (1.9) one easily obtains that $(u_n)^{1/n}$ tends to $\rho \leqslant 1$ as n approaches infinity. Evidently (1.10) implies that $\rho = 1$ if (1.5) holds.

Example 1. Let the state space be $S = \{0, 1, \ldots\}$ and define the transition probabilities as follows: $P(0,0) = \alpha_0$, $P(0,1) = \beta_0$, $P(0,2) = 1-(\alpha_0 + \beta_0)$; for $n \geqslant 1$, $P(2n, 2n-1) = \alpha_n$, $P(2n, 2n+1) = \beta_n$, $P(2n, 2n+2) = 1-(\alpha_n + \beta_n)$; $P(1, 0) = 1$, and for $n \geqslant 1$, $P(2n+1, 2n-1) = 1$; in all other cases $P(i,j) = 0$. Here the α_n and β_n are free to be determined, $0 \leqslant \alpha_n$, $0 \leqslant \beta_n$, $\alpha_n + \beta_n \leqslant 1$. If for some n $\alpha_n + \beta_n = 1$ cut down on the state space by deleting all states corresponding to integers greater than $2n+1$. It is clear that given (f_n) satisfying (1.3) and (1.4) one can choose the α_n and β_n so that (1.2) holds.

For the class of chains of this example it is clear that if u_{n+1}/u_n does not converge then $P^n(0,2)/P^n(0,0)$ does not converge. Note that if (R) and (A) hold, that is, (1.5) and (1.6) hold, u_{n+1}/u_n need not converge.

Indeed let (g_n), $n = 1, 2, \ldots$ be a sequence of positive numbers, $\sum_{n=1}^{\infty} g_n = 1$. Let $1 = n_1 < n_2 < n_3 \ldots$ be a sequence of positive integers and set $f_n = 0$ if $n \notin \{n_1, n_2 \ldots\}$, $f_{n_i} = g_i$. For suitable choice of the n_i u_{n+1}/u_n will not converge. A suitable choice will be defined by induction. Suppose $n_1, n_2 \ldots n_k$ defined. Let $\overline{f}_n = 0$ for $n \notin \{n_1, n_2 \ldots n_k\}$, $\overline{f}_{n_i} = g_i$, $1 \leq i \leq k$. Then $\sum_{n=1}^{\infty} \overline{f}_n < \infty$ and so the corresponding $\overline{u}_n$ approach zero. Let n_{k+1} be chosen so that $\overline{u}_{n_{k+1}-1} \leq 2^{-k} g_{k+1}$. Observe that for $n < n_{k+1}$ $u_n = \overline{u}_n$ so one will have

$$u_{n_{k+1}} \geq f_{n_{k+1}} = g_{k+1} \geq 2^k \overline{u}_{n_{k+1}-1} = 2^k u_{n_{k+1}-1}$$

Thus by suitable choice of a_n and β_n we obtain chains which satisfy (A) and (R), but $P^n(0,1)/P^n(0,0)$ does not converge. Note that the class of chains of this example has no large jumps: $P(i,j) = 0$ if $|i-j| > 2$.

For chains satisfying (DR) we will say that the *strong ratio limit property* holds if

$$(SRLP) \quad \lim_{n \to \infty} \frac{P^{n+m}(i,j)}{P^n(a,k)} = \frac{\pi(j)}{\pi(k)}, \qquad i, j, a, k \in S, \qquad m = 0, \pm 1, \ldots.$$

Despite the counter example, it will be seen that under (DR) and (A), it is more the rule than the exception that $(SRLP)$ holds.

Convention: When not explicitly indicated, limits will be taken as n tends to infinity. The symbol $\sim$ is written to indicate that the

two terms connected by this symbols asymptotically have the ratio
1. Thus $u_{n+1}/u_n \to 1$ and $u_{n+1} \sim u_n$ both mean

$$\lim_{n \to 0} \frac{u_{n+1}}{u_n} = 1.$$

The following proposition will be useful below.

PROPOSITION 1.2. *Let the f_k satisfy (1.3), (1.5) and (1.6), and let the u_n be defined by (1.7). Let d be a positive integer. Then* $\limsup u_{(n+1)d}/u_n \leqslant 1$ *implies that* $\lim u_{n+1}/u_n = 1$.

Proof. Define

$$a_j = \liminf u_{nd-j}/u_{nd}, \qquad b_j = \limsup u_{nd-j}/u_n.$$

and note that the hypothesis of the theorem implies:

(1.14) $\quad a_{d+j} = \liminf u_{nd-d-j}/u_{nd}$

$\qquad \geqslant \liminf u_{nd-d-j}/u_{nd-j} \cdot \liminf u_{nd-j}/u_{nd} \geqslant a_j.$

Write the renewal equation (1.7) for u_{nd-j}, divide by u_n and use Fatou's lemma to obtain

(1.15) $$a_j \geqslant \sum_{k=1}^{\infty} f_k \, a_{j+k}.$$

From (1.14) it follows that a_i attains its minimum value c for some index j', $0 \leqslant j' < d$. Equation (1.15) implies $a_{j'+k} = c$ for all k such that $f_k > 0$. Iteration and use of (1.6) gives $a_j = c$ for all sufficiently large c, and hence by (1.14) $a_j = c$ for all j. Since $a_0 = 1$, $a_j = 1$ for all j.

Fix i and let n' increase to infinity so that $u_{n'd-i}/u_{n'}$ tends to $\limsup u_{nd-i}/u_n$. In place of (1.15) one obtains

$$b_j \geq \sum_{k=1}^{\infty} f_k a_{j+k} + f_{i-j}(b_i - a_i),$$

and hence

$$(b_j - 1) \geq f_{i-j}(b_i - 1), \qquad 0 \leq j < i.$$

Since $b_0 = 1$, one has $b_i = 1$ whenever $f_i > 0$, and again by iteration $b_i = 1$ for all sufficiently large values of i. So for all sufficiently large j, u_{nd-j}/u_{nd} tends to 1. This obviously implies that u_{n+1}/u_n tends to 1.

THEOREM 1.1. *Consider a Markov chain satisfying* (DR) *and* (A). *Let* $u_n = P^n(0, 0)$. *If for some positive integer* d
$\lim \sup u_{(n+1)d}/u_{nd} \leq 1$ *then* (SRLP) *holds.*

Proof. Assume the hypothesis. According to the preceding Proposition this implies $u_{n+1} \sim u_n$. So the theorem need only be proved for the case $d = 1$. For this case the result follows from (iii) of Theorem 1.2 below, since under (DR) the measures $\varphi = \epsilon_i$, $i \in S$ are absolutely continuous with respect to π.

Consider a Markov chain on $(S, \mathcal{B})$ satisfying (R) and (A). The hypothesis

$$\lim_{n \to \infty} \frac{P^{n+1}(0,0)}{P^n(0\ 0)} = 1$$

is not useful for general state space. In its place one may assume that there exists a set $B \in \mathcal{B}$ with $\pi(B) > 0$ such that

(1.16) $\quad \lim_{n \to \infty} \int P^{n+1}(w, dy) f(y)/P^n(z, B) = \int f(y) \pi(dy)/\pi(B)$

for every $w \in B$, $z \in B$ and bounded measurable function f vanishing on $S-B$. It is not hard to show that this requirement is equivalent to demanding that (1.16) hold for all f of the form $f = I_A$, where

$A \subseteq B$, $A \in B$. Another condition required below is that a set A be well behaved with respect to B in the sense that for some M and some j

$$(1.17) \quad {}_B P^n(x, A) \leq M \cdot \sum_{k=1}^{j} {}_B P^{n+k}(x, B), \qquad x \in S, \qquad n = 0, 1, \ldots.$$

Note in the first place that if

$$(1.18) \quad \inf_{y \in A} \sum_{k=1}^{j} {}_B P^k(y, B) > 0 \text{ for some } j,$$

then (1.17) automatically holds. It is also easy to see that if

$$(1.19) \quad A \subseteq C \text{ for some set } C \text{ which is a } C\text{-set}$$

then (1.18) and (1.17) will be satisfied for any B such that $\pi(B) > 0$. Condition (1.19) has the advantage of being independent of B.

THEOREM 1.2. *Let the Markov chain on* $(S, \mathcal{B})$ *satisfy* (R) *and* (A). *Let* $B \in \mathcal{B}$ *and suppose* $\pi(B) > 0$ *and* (1.16) *holds for all* $z \in B$, $w \in B$ *and all bounded measurable functions* f *vanishing on* $S-B$, *then*

(i) $\liminf_{n \to \infty} P^{n+m}(x, A)/P^n(z, B) \geq \pi(A)/\pi(B)$, $m = 0, \pm 1, \ldots$,
    $z \in B$, $x \in S$, $A \in \mathcal{B}$.

(ii) $\lim_{n \to \infty} P^{n+m}(x, A)/P^n(z, B) = \pi(A)/\pi(B)$, $m = 0, \pm 1, \ldots$, $z \in B$,
    provided $x \in S$ *satisfies* $\limsup_{n \to \infty} P^{n+i}(x, B)/P^n(w, B) \leq 1$ *for some integer* i *and some* $w \in B$ *(equivalently by* (1.13) *every integer* i *and every* $w \in B$) *and provided* $A \in \mathcal{B}$ *satisfies* (1.17).

(iii) *For any finite measure* φ *on* (S, B) *which is absolutely continuous with respect to* π *and each* $A \in \mathcal{B}$ *satisfying* (1.17)

$$\lim_{n \to \infty} \varphi\{x : |P^{n+m}(x,A)/P^n(z,B) - \pi(A)/\pi(B)| > \epsilon\} = 0$$

for each $z \in B$, $\epsilon > 0$, $m = 0, \pm 1, \ldots$.

Proof. Iterating (1.16) with $f = I_B$ gives

(1.20) $$\lim_{n \to \infty} P^{n+m}(w,B)/P^n(z,B) = 1,$$

$$m = 0, \pm 1, \ldots, \qquad w \in B, \qquad z \in B.$$

It therefore suffices to prove (i) for $m = 0$. If the chain moves from x to A in n steps it either fails to enter B during these n steps or enters B for the first time at time v and for the last time at some time n-k, and one obtains for $n \geqslant N \geqslant 0$,

(1.21) $$P^n(x,A) = \sum_{v=1}^{n-1} \int_B {}_B P^v(x, dy)$$

$$\times \sum_{k=1}^{n-v} \int_B P^{n-v-k}(y, dz) {}_B P^k(z,A) + {}_B P^n(x,A)$$

$$= s_{n,N}(x,A) + r_{n,N}(x,A),$$

where

(1.22) $$s_{n,N}(x,A) = \sum_{v=1}^{N} \int_B {}_B P^v(x, dy)$$

$$\times \sum_{k=1}^{N} \int_B P^{n-v-k}(y, dz) {}_B P^k(z,A).$$

Let $z \in B$. In (1.22) divide through by $P^n(z,B)$ and let n approach infinity, using (1.16) to conclude:

(1.23) $$\lim_{n \to \infty} s_{n,N}(x,A) \cdot (P^n(z,B))^{-1}$$

$$= \sum_{v=1}^{N} {}_B P^v(x,B) \cdot \sum_{k=1}^{N} \int_B {}_B P^k(z,A)\, \pi(dz) \cdot (\pi(B))^{-1}.$$

As N tends to infinity the first factor on the right side of (1.23) tends to 1, whereas the second factor tends to $\pi(A)$, as was shown in Theorem 7.2 of Chapter 1 (see especially (7.2)). It follows that

$$(1.24) \quad \lim_{N \to \infty} \lim_{n \to \infty} s_{n,N}(x,A) \cdot (P^n(z,B))^{-1} = \pi(A)/\pi(B),$$

and this together with (1.21) proves (i). Note that equality will hold in (i), with $m = 0$, fixed $z \in B$, if and only if

$$(1.25) \quad \lim_{N \to \infty} \lim_{n \to \infty} r_{n,N}(x,A) \cdot (P^n(z,B))^{-1} = 0,$$
$$z \in B, \quad m = 0, \pm 1, \ldots.$$

Turning to the proof of (ii) it again suffices to consider $m = 0$, $z \in B$ fixed. Suppose now that x satisfies the first proviso of (ii). The equivalence noted in parentheses obviously follows from (1.20). Let $A \in \mathcal{B}$ satisfy (1.17). It must be shown that (1.25) is satisfied. Equality holds in (i) when A is replaced by B, because of the assumption on x, and so the condition corresponding to (1.25) must hold

$$(1.26) \quad \lim_{N \to \infty} \lim_{n \to \infty} r_{n,N}(x,B) \cdot (P^n(z,B))^{-1} = 0.$$

Condition (1.17) implies

$$r_{n,N}(x,A) \leqslant M \cdot \sum_{k=1}^{j} r_{n+k,N}(x,B),$$

and this together with (1.26) implies the truth of (1.25).

It needs to be noted that the proof of (ii) shows that if in (1.16) and the first proviso of (ii) n is replaced by n', where n' increases to infinity along some subsequence of the positive integers, the conclusion of (ii) holds with n' in place of n. This remark will

allow us to reduce (iii) to the special case $A = B$, and $m = 0$. To
see this let $z \in B$, and set

$$g_n(x) = P^n(x,B)/P^n(z,B).$$

Suppose then that every subsequence $(g_{n'})$ of (g_n) has a further
subsequence $(g_{n''})$ converging to 1 π-a.e.; note that this is
equivalent to convergence in measure for each finite measure φ
absolutely continuous with respect to π. Then for any x such that
$g_{n''}(x)$ approaches 1, (ii) can be applied with n'' in place of n.

Procceding then to the case $A = B$, $m = 0$, let $z \in B$ and write

(1.27) $$1 = \frac{P^n(z,B)}{P^n(z,B)} = \int \frac{P^{n-k}(y,B)}{P^n(z,B)} \cdot P^k(z,dy).$$

As n tends to infinity, for each k the limit inferior of the integrand
in (1.27) is at least 1 by (i) and it follows that the integrand must
converge to 1 in $P^k(z,\cdot)$-measure. Using also (1.20) one sees that
g converges to 1 in $P^k(z,\cdot)$ measure, $k = 1, 2, \ldots$. A diagonal
argument shows that each subsequence $(g_{n'})$ of (g_n) has a further
subsequence $(g_{n''})$ converging to 1 $P^k(z,\cdot)$-a.e. for all $k > 0$, and
(R) implies that $(g_{n''})$ converges to 1 π-a.e., as had to be shown.

Example 1.2. Let $Y_1, Y_2, \ldots$ be a sequence of independent,
identically distributed, real random variables. Let X_0 be a
non-negative random variable independent of the Y_k and define

$$X_n = (X_{n-1} + Y_n)^+, \qquad n = 1, 2, \ldots,$$

the notation x^+ standing for max$(x,0)$. Then $(X_n, n = 0, 1, \ldots)$ is a
Markov process on $[0,\infty)$ with transition probability $P(x,E)$, say.
(This chain differs from random walk on the line, for which see
next example, in that 0 acts as a barrier for the former but not the

latter.) We are interested in the case $L(x,\{0\}) = 1$ for all $x \geqslant 0$, i.e. from any starting point $\{0\}$ is sure to be entered. It is known that this is equivalent to

$$\sum_{n=1}^{\infty} n^{-1} P[S_n < 0] = \infty,$$

where

$$S_n = \sum_{k=1}^{n} Y_k,$$

but this fact will not be needed here. Evidently (X_n) is φ-recurrent for any φ such that $\varphi(\{0\}) > 0$ and so it has an invariant measure π. Let $M_n = \max\{S_k : 1 \leqslant k \leqslant n\}$. It turns out that if $I_y = [0,y]$ then

(1.28) $\qquad P^n(x,I_y) = P[M_{n-1}^+ \in I_y, (S_n + x)^+ \in I_y].$

To see this note that induction on n shows

$$X_n = \max\{0, S_n + X_0, \max(Y_n, Y_n + Y_{n-1}, \ldots, Y_n + Y_{n-1} + \ldots + Y_2)\},$$

and therefore

$$P^n(x, I_y) = P[(S_n + x)^+ \in I_y,$$
$$\max(Y_n, Y_n + Y_{n-1}, \ldots, Y_n + Y_{n-1} + \ldots + Y_2) \in I_y],$$

and since the Y_n are independent and identically distributed the right side of the last equality equals that of (1.28). From (1.28) it follows at once that

(1.29) $\qquad P^n(0, I_y) \leqslant P^m(0, I_y) \quad \text{for } n \geqslant m,$

(1.30) $\qquad P^n(x, I_y) \leqslant P^n(0, I_y) \quad \text{for } x \geqslant 0.$

These relations will be required only for $y = 0$. Put $u_n = P^n(0, \{0\})$. By (1.29) (u_n) is monotone and so Proposition 1.2 applies to show that u_{n+1}/u_n converges to 1 as n tends to infinity. With $B = \{0\}$ this implies (1.16), and (1.30) becomes $P^n(x,B) \leq P^n(0,B)$ so that every $x \in [0,\infty)$ satisfies the proviso in (ii) of Theorem 1.2. Finally, every bounded Borel set A included in $[0,\infty)$ is easily seen to satisfy (1.18), hence also (1.17), and so for these sets the conclusion of (ii) of Theorem 1.2 will hold.

Example 1.3. Modify the previous example by dropping the requirement that X_0 be non-negative. Let $Z_0 = X_0$, $Z_n = S_n + X_0$ $n = 1, 2, \ldots$. Then $(Z_n, n = 0, 1, \ldots)$ is a Markov chain, *random walk* on the line. This is an interesting and important example, but it cannot be effectively studied without using the topology and group structure of the line. In particular, one must exploit the fundamental property of random walk, the fact that the transition probabilities satisfy $P(x,A) = P(x+y, A+y)$. In the study of these chains the natural recurrence condition is not (R), but the weaker one that $Q(x,E) = 1$ for every x and every open set E.

Returning to the situation of a chain satisfying (R) and (A), and assuming $\mathcal{B}$ separable, when can one assert that

$$(1.31) \qquad \lim_{n \to \infty} \frac{\int P^{n+m}(x,A)\pi(\mathrm{d}x)}{\int P^n(x,E)\nu(\mathrm{d}x)} = \frac{\pi(A)}{\pi(E)}, \qquad m = 0, \pm 1, \ldots,$$

for every pair of initial probability distributions μ and ν and all C-sets A and E? This requirement is indeed pretty restrictive, as can be seen by the precise characterization of these chains given in the next theorem.

THEOREM 1.3. *Let the Markov chain on $(S, \mathcal{B})$ satisfy (R) and (A) and suppose that $\mathcal{B}$ is separable. Then (1.31) holds for all pairs of initial probability distributions μ and ν and all C-sets A and E if*

and only if there exists a C-set B such that (1.16) holds for every
$z \in B$, $w \in B$ and bounded measurable function f vanishing on S-B
and there exist $z \in B$ and positive integers M and N such that

(1.32) $P^{n+1}(x, B)/P^n(z, B) \leqslant M,$ $x \in S,$ $n \geqslant N.$

Proof. To begin with the 'only if' part, it is evident that there
must exist a C-set B satisfying (1.16) if (1.31) is to be satisfied.
If the condition involving (1.32) is violated, then for any $z \in B$
there exist integers n_k and $x_k \in S$ such that

$$P^{n_k+1}(x_k, B) \geqslant 2^{k+1} P^{n_k}(z, B), \qquad k = 1, 2, \ldots.$$

Let μ be the probability measure concentrated on $\{x_k : k = 1, 2, \ldots\}$
with $\mu(\{x\}) = 2^{-k}$, $k = 1, 2, \ldots$ and let $\nu = \epsilon_z$, to obtain a counter
example to (1.31) with $A = E = B$, $m = 1$.

For the converse implication assume B is a C-set such that
(1.16) holds for all $z \in B$, $w \in B$ and bounded measurable f vanishing
on S-B, and let $z \in B$, M and N be such that (1.32) holds. It is
clearly enough to show that (1.31) holds for all initial probabilities
μ, all C-sets A and $\nu = \epsilon_z$, $E = B$. The first step will be to show
that this is true in the special case $A = B$, $m = 1$ and $\mu = \epsilon_x$, where
$x \in S$.

Now for $1 \leqslant k \leqslant n$

$$\frac{P^{n+1}(x, B) - P^n(z, B)}{P^n(z, B)}$$
$$= \int \left\{ [P^{k+1}(x, dy) - P^k(z, dy)] \cdot \frac{P^{n-k}(y, B)}{P^{n-k-1}(z, B)} \right\} \cdot \frac{P^{n-k-1}(z, B)}{P^n(z, B)}.$$

For $n \geqslant N + k$ the second factor in the integrand is bounded by M,
and as n tends to infinity the factor multiplying the integral goes to
1 by (1.16).

Therefore

$$\limsup_{n\to\infty} \left| \frac{P^{n+1}(x,B) - P^n(z,B)}{P^n(z,B)} \right| \leq M \cdot ||(\epsilon_x P - \epsilon_z)P^k||,$$

and by the Corollary to Theorem 5.1 of Chapter 1 the last term tends to zero as k approaches infinity. This completes the proof for the case $A = B$, $m = 1$, $\mu = \epsilon_x$.

So every $x \in S$ satisfies the proviso in (ii) of Theorem 1.2, and therefore the conclusion of (ii) holds for any A satisfying (1.17), and as already remarked (1.19) implies (1.17). To complete the proof of the theorem it will be shown that one may integrate the equality expressing the conclusion of (ii) with respect to μ and interchange the order of integration and passage to the limit. Considering then the case $m = 1$, which suffices, it will be enough to show that $| P^{n+1}(x,A)/P^n(z,B) |$ is bounded uniformly in n and x, where A is a fixed C-set. The fact that A is a C-set is easily seen to imply the existence of an integer n_0 and a constant K such that

$$P^n(x,A) \leq K P^{n+n_0}(x,B), \qquad n = 0, 1, \ldots, \qquad x \in S,$$

and this relation, together with the hypothesis (1.32) and the assumption that B satisfies (1.16), gives the desired bound for $| P^{n+1}(x,A)/P^n(z,B) |$. This then concludes the proof of (1.31) for arbitrary C-sets A, arbitrary μ, $m = 1$ and $B = E$, $\mu = \epsilon_z$, from which the general case follows immediately.

We conclude this section with a stochastic individual ratio limit theorem—the only stochastic convergence result in these notes.

THEOREM 1.4. *Consider a chain satisfying (D). Let a, b $\in$ S, and let m be a non-negative integer. Let*

$$h_n(x) = \frac{P^{n+m}(a,x)}{P^n(b,x)}, \qquad n = 0, 1, \ldots,$$

the fraction to be interpreted as zero when the denominator vanishes. Then $h_n(X_n)$ converges P_b-a.s. If the chain satisfies (DR) and (A) $h_n(X_n)$ converges to 1 P_μ-a.s. for every initial distribution μ.

Proof. Let the Markov chain be started with initial distribution ϵ_b. It is to be shown that $(h_n(X_n), n = \ldots 2, 1, 0)$ is a supermartingale. Indeed P_b-a.s.

$$(1.33) \quad E_b[h(X_n) \mid X_{n+1}, X_{n+1}, \ldots] = E_b[h_n(X_n) \mid X_{n+1}] \leqslant h_{n+1}(X_{n+1}).$$

The equality is just a consequence of the Markov property (see Loeve [1955], p.563). To show the inequality in (1.33) holds P_b-a.s. it will suffice to show that $E_b[h_n(X_n) \mid X_{n+1} = x] \leqslant h_{n+1}(x)$ for any x such that $P_b[X_{n+1} = x] > 0$. Let $S_n = \{y : P^n(b,y) > 0\}$, $x \in S_{n+1}$. Note

$$(1.34) \quad E_b[h_n(X_n)] = \sum_{y \in S_n} P^n(b,y) \left(\frac{P^{n-m}(a,y)}{P^n(b,y)} \right)$$

$$= \sum_{y \in S_n} P^{n-m}(a,y) \leqslant 1,$$

and further

$$(1.35) \quad E_b[h_n(X_n) \mid X_{n+1} = x] = \sum_y h_n(y) P_b[X_n = y \mid X_{n+1} = x]$$

$$= \sum_{y \in S_n} \frac{P^{n+m}(a,y)}{P^n(b,y)} \cdot \frac{P^n(b,y)P(y,x)}{P^{n+1}(b,x)}$$

$$= \sum_{y \in S_n} \frac{P^{n+m}(a,y)P(y,x)}{P^{n+1}(b,x)} \leqslant \frac{P^{n+1-m}(a,x)}{P^{n+1}(b,x)}$$

$$= h_{n+1}(x).$$

By (1.34) and (1.35) the backward super-martingale convergence theorem applies to give $h_n(X_n)$ converges to a finite limit H-a.s.

In view of (1.35) the expectation in the first member of (1.34) is non-increasing as a function of n. If the chain satisfies (DR) and (A), S_n converges to S, and the last sum in (1.34) approaches 1. Hence in that case $(\ldots, h_2(X_2), h_1(X_1), h_0(X_0))$ is a martingale, $E[h_n(X_n)] = 1$. The backward martingale convergence theorem implies that the limit H also satisfies $E[H] = 1$. Writing $h((x,n)) = h_n(x)$, one has $H = \lim h((X_n, n))$, so that H is evidently equal a.s. to an invariant random variable for the space-time process (X_n, n), $n = 0, 1, \ldots$. For each N, $H \wedge N$ is a bounded invariant random variable of the space-time process, hence constant by Chapter 1, Theorem 5.1 and Proposition 4.3. Since $H \wedge N$ converges to H as N approaches infinity H is a.s. constant. As $E[H] = 1$, $H \equiv 1$-a.s. Obvious manipulation then gives $P^{n+m}(a, X_n)/P(c, X_n) \to 1$ P_b-a.s. for all a, c, $b \in S$. This implies that the indicated convergence takes place with probability one for any initial probability μ.

Remark 1.1. The result extends to general state space $(S, \mathcal{B})$ with $\mathcal{B}$ separable. Then h_n must be defined as the Radon–Nikodym derivative of the $P^n(b, \cdot)$-absolutely-continuous part of $P^{n+m}(a, \cdot)$ with respect to $P^n(b, \cdot)$.

Remark 1.2. Consider Theorem 1.3 for chains satisfying (DR) and (A). Let $a = b = 0 \in S$. Let T_1, T_2, ... be the successive return times to 0, $u_n = P_{00}^n$. The theorem implies $u_{T_n+m}/u_{T_n} \to 1$-a.s., $m = 0, \pm 1, \ldots$. From this one gets the following purely analytic fact. Starting with any sequence (f_k) satisfying (1.3), (1.5), and (1.6) and defining u_n by (1.7) there always exists a sequence n' increasing to infinity such that $u_{n'+m}/u_{n'} \to 1$ as $n' \to \infty$, $m = 0, \pm 1, \ldots$.

2. *Probabilistic conditions for* SRLP. Consider chains satisfying (DR) and (A). We know that they do not necessarily enjoy $(SRLP)$ (see previous section for definition, right after Example 1.1). In looking for conditions for $(SRLP)$ we can usefully distinguish between two types: conditions on the structure of the Markov chain, and conditions on the first recurrence times to some specific state 0. The first kind of condition will be discussed in this section. The second kind of condition will be treated in Section 3 and 4; note that according to Theorem 1.1 $(SRLP)$ under our assumption is equivalent to $u_n \sim u_{n+1}$. An investigation concerning conditions on the f_k implying $u_{n+1} \sim u_n$ simply involves the renewal equation. Of course the two approaches are connected. The following example is illustrative.

Example 2.1. Consider a chain satisfying (DR) and (A). Suppose for all i and j $\pi(i)P(i,j) = \pi(j)P(j,i)$. A chain satisfying this condition is called *reversible*. It is known that in this case $u_n = \int_{-1}^{1} s^n \, d\psi(s)$, where ψ is a bounded measure on $[-1, 1]$. Therefore (u_{2n}) is non-increasing as a function of n and $(SRLP)$ follows from Theorem 1.1. The condition of reversibility is equivalent to the following: for any given finite sequence of states $i_0, i_1, \ldots, i_n$ $P(i_0, i_1)P(i_1, i_2)\ldots P(i_{n-1}, i_n)P(i_n, i_0) = P(i_0, i_n)P(i_n, i_{n-1})\ldots P(i_2, i_1)P(i_1, i_0)$. One class of instances is provided by chains having the integers for state space and satisfying $P(i,j) = 0$ for $|i-j| > 1$.

The next proposition is required for the proof of Theorem 2.1.

PROPOSITION 2.1. *Let (X_i) $i = 1, 2, \ldots$ be a sequence of independent identically distributed random variables, each having a distribution:* $P[X_i = 1] = p$, $P[X_i = 0] = 1-p$. *Let* $S_n = \sum_{i=0}^{n} X_i$. *For every* $\epsilon > 0$, $P[|S_n - np| > \epsilon n]$ *approaches zero exponentially fast.*

Proof. It will suffice to show that $P[(S_n - np) > \epsilon n]$ goes to zero at an exponential rate; the assertion with absolute values then follows by considering variables $Y_i = 1 - X_i$ in place of the X_i.

For a real one has

$$e^{a\epsilon n} P[(S_n - np) > \epsilon n] \leq E[\exp\{a(S_n - np)\}] = (E[\exp\{a(X_1 - p)\}])^n$$

$$= (p\, e^{a(1-p)} + (1-p)\, e^{-ap})^n$$

As a tends to zero the expression in parentheses in the last member equals $1 + 1/2\, p(1-p)a^2 + 0(a^2)$, and there exists a constant k_p such that for all a sufficiently close to zero this expression is less than $1 + k_p a^2$, hence less than $\exp\{k_p a^2\}$. Thus $P[(S_n - np) > \epsilon n] \leq [\exp\{k_p a^2 - a\epsilon\}]^n$, and when $\epsilon > 0$ and a is small enough the expression in braces will be negative.

THEOREM 2.1. *Consider a chain satisfying (DR) and (A). Suppose m is a positive integer, $\epsilon > 0$ and $\sum_{k=1}^{m} P^k(i,i) > \epsilon > 0$ holds for all $i \in S$. Then (SRLP) holds.*

Proof. Evidently for each i there exists $m_i \leq m$ such that $P^{m_i}(i,i) > \epsilon/m$. Therefore $P^{m!}(i,i) \geq (\epsilon/m)^{m!/m_i} \geq (\epsilon/m)^{m!} > 0$. Below it will be shown that $P(i,i) \geq \Delta > 0$ for all i implies $\limsup u_{n+1}/u_n < 1$. Therefore $P^{m!}(i,i) \geq \Delta > 0$ implies $\limsup u_{m!(n+1)}/u_{m!n} \leq 1$, and by Theorem 1.1 (SRLP) follows.

Assume then that $P(i,i) > \Delta > 0$ for all i. Let $P^*(i,j) = (P(i,j) - \Delta \delta(i,j))/(1-\Delta)$, where $\delta(i,j) = 0$ for $i \neq j$, $\delta(i,i) = 1$. Consider a particle moving on S according to the following rule: at time 0 it is at 0, at time $n+1$ a coin is flipped which has probability of coming up heads (tails) equal to $1-\Delta$ (Δ), $n = 0, 1, \ldots$. If heads appears the particle makes a transition according to the transition probability matrix P^*; if tails come up the particle makes no transition. Let X_n be the position of the particle at time n. Then (X_n) is a Markov chain with initial distribution ϵ_0 and transition probability matrix P. Let N_n be the number of transitions up to time n, and let Z_k be the position of the particle after k transitions. Observe that $P[N_n = k] = b(k; n, 1-\Delta)$, where $b(k; n, v)$ is the probability of k successes in n Bernoulli trials, the probability of success at each trial being equal to v. Note now that the sequences $Z_0, Z_1, \ldots$ and $N_0, N_1, \ldots$ are independent. Therefore

$$(2.1) \quad u_n = P[X_n = 0] = \sum_k P[X_n = 0, N_n = k]$$

$$= \sum_k P[Z_k = 0, N_n = k]$$

$$= \sum_k P[Z_k = 0] b(k; n, 1-\Delta).$$

Let $\delta \in (0, \Delta/2)$. Write Σ' for the sum over all k satisfying $|k-(1-\Delta)n| < \delta n$, and Σ'' for the sum over all k such that $|k-(1-\Delta)n| \geqslant \delta n$. From (2.1) one obtains

$$(2.2) \quad \frac{u_{n+1}}{u_n} = \frac{\Sigma' P[Z_k = 0] b(k; n+1, 1-\Delta)}{\sum_k P[Z_k = 0] b(k; n, 1-\Delta)} + \frac{\Sigma'' P[Z_n = 0] b(k; n+1, 1-\Delta)}{u_n}.$$

The numerator of the last term approaches zero at an exponental rate, by Proposition 2.1, while the denominator does not go to zero that rapidly according to (1.13). Hence the term approaches zero. On the other hand one verifies directly that

$$\frac{b(k; n+1, 1-\Delta)}{b(k; n, 1-\Delta)} \leqslant 1 + \frac{2\delta}{\Delta} + \frac{2}{n}$$

for all k in the range of the first summation. Thus the limit superior of the first term on the right of (2.2) is at most $1 + 2\delta/\Delta$, and since δ is arbitrary $\limsup u_{n+1}/u_n \leqslant 1$. The theorem is proved.

In the proof of the theorem, relation (2.1) can be generalized to

(2.3) $$P^n(i,j) = \sum_k P^{*k}(i,j)\, b(k;n, 1-\Delta).$$

Now P^* is the transition probability of a chain satisfying (DR), but not necessarily (A). Conversely, starting with such a P^* the relation $P^*(i,j) = (P(i,j) - \Delta\delta(i,j))/(1-\Delta)$ determines a transition probability P satisfying the hypotheses of Theorem 2.1, indeed $P(i, i) \geq \Delta$. We therefore can draw the following conclusion:

COROLLARY. *If P^* is the transition probability matrix of a chain satisfying (DR) and $b(k; n, v)$ denotes the probability of k successes in n Beruoulli trials with probability of success of each trial being v, then*

$$\lim_{n \to \infty} \frac{\sum_k P^{*k}(x,j) b(k; n+m, 1-\Delta)}{\sum_k P^{*k}(y,i) b(k; n, 1-\Delta)} = \frac{\pi(j)}{\pi(i)},\ 0 < \Delta < 1,\ x,\ j,\ y,\ i \in S.$$

It seems likely that the foregoing theorem and its corollary have valid generalizations for general state space. Presumably the right condition for the theorem is that every x is in some C-set C such that $p^n(y, z) > \epsilon > 0$ for some n and ϵ and all $y \in C$, $z \in C$ and n and ϵ can be chosen fixed independently of x. The corollary should involve no condition other than φ-recurrence.

The next proof will require the fact that if r is a positive integer

(2.4) $$\lim_{n \to \infty} r^{-1}(u_{n-1} + u_{n-2} + \ldots u_{n-r})/u_n = 1 \text{ implies } u_{n+1} \sim u_n.$$

This is an immediate consequence of Theorem 3.1 below, obtained by letting $a_i = r^{-1}$, $i = 1, 2, \ldots r$, $a_i = 0$ for $i > r$.

The next theorem gives a set of conditions which are equivalent to $(SRLP)$. For its statement we introduce some notation. Let $A \in \mathcal{B}_0$ and define:

$$V_A = \min\{k > 0 : X_k \in A\},$$

$$Y_n(A) = \begin{cases} n - \max\{k \leq n : X_k \in A\} & V_A \leq n \\ n & V_A > n \end{cases},$$

$$Z_n(A) = \min\{k > n : X_k \in A\} - n,$$

$$T_n(A) = Y_n(A) + Z_n(A).$$

THEOREM 2.2. *Assume (DR) and (A). If (SRLP) holds then for any finite non-void set of states A, any positive integer r such that $\sum_{y \in A} P_y[V_A = r] > 0$, and any k such that $0 \leq k < r$ and any $x \in A$,*

(2.5) $\lim_{n \to \infty} P_x[Y_n(A) = k | T_n(A) = r]$
$$= \lim_{n \to \infty} P_x[Z_n(A) = r-k | T_n(A) = r] = r^{-1}.$$

Conversely, if (*) holds for the special case $A = \{0\}$, $x = 0$, any r such that $f_r > 0$, and $k = 0$, then (SRLP) holds.

Proof. Evidently,

(2.6) $P_x[Y_n(A) = k, T_n(A) = r]$
$$= \sum_{y \in A} P^{n-k}(x,y) P_y[V_A = r], \qquad 0 \leq k \leq (r-1) \wedge n,$$

and therefore

(2.7) $$P_x[T_n(A) = r] = \sum_{k=0}^{(r-1)\wedge n} \sum_{y \in A} P^{n-k}(x,y) P_y[V_A = r].$$

Dividing the preceding relations by u_n and assuming (SRLP) one obtains

$$u_n^{-1} P_x[Y_n(A) = k, T_n(A) = r] \sim \sum_{y \in A} \frac{\pi(y)}{\pi(0)} P_y[V_A = r],$$

and also

$$u_n^{-1} P_x[T_n(A) = r] \sim \sum_{y \in A} r \frac{\pi(y)}{\pi(0)} P_y[V_A = r],$$

and (2.5) follows.

For the converse, assume that r is such that $P_0[V_0 = r] = f_r > 0$ and (2.5) holds with $x = 0$, $A = \{0\}$, $k = 0$. By (2.6) and (2.7) then, for $n \geq r$

$$P_0[Y_n(0) = 0 | T_n(0) = r] = \frac{P^n(0,0) P_0[V_0 = r]}{\sum_{k=0}^{r-1} P^{n-k}(0,0) P_0[V_0 = r]}$$

$$= \frac{u_n}{\sum_{k=0}^{r-1} u_{n-k}} \to \frac{1}{r},$$

and thus $u_{n+1} \sim u_n$ by (2.4). Finally (SRLP) follows by Theorem 1.1.

3. *Necessary and sufficient conditions for $u_{n+1} \sim u_n$.*
Throughout this and the next section (f_k), $k = 1, 2, \ldots$ is to be a sequence satisfying (1.3), (1.5) and (1.6), and (u_n), $n = 0, 1, \ldots$ is to be the sequence determined by the renewal equation (1.7).

Various conditions which are both necessary and sufficient for $u_{n+1} \sim u_n$ are known. Theorem 3.1 below and the first corollary to that theorem provide instances. None of these conditions is satisfactory, because the sequence (u_n) is itself involved. Nevertheless, these results have intrinsic interest, and, as will be seen in the next section, they are useful for obtaining sufficient conditions on the (f_k) for $u_{n+1} \sim u_n$.

Definition 3.1. Let $m = \liminf_{n \to \infty} (u_{n+1}/u_n)$, $M = \limsup_{n \to \infty} (u_{n+1}/u_n)$.

Evidently $0 \leqslant m \leqslant M \leqslant \infty$. Dividing the renewal equation by u_{n-1} and using Fatou's lemma one obtains

$$(3.1) \qquad m = \liminf_{n \to \infty} \frac{u_n}{u_{n-1}} \geqslant M \sum_{k=1}^{\infty} f_k M^{-k} > 0 \text{ if } M < \infty.$$

Relation (3.1) again makes evident that to prove $u_{n+1} \sim u_n$ it suffices to show $M \leqslant 1$.

PROPOSITION 3.1. *If for some $n_1 \geqslant 1$*

$$(3.2) \qquad u_n = O(u_{n-1} + u_{n-2} + \cdots u_{n-n_1}),$$

then for every integer k

$$(3.3) \qquad u_n = O(u_{n+k}),$$

and in particular it follows from the cases $k = -1$ and $k = 1$ that

$$(3.4) \qquad 0 < m \leqslant M < \infty.$$

Proof. According to (1.9) $u_{n+k} \geqslant u_n u_k$, and since, by (1.11) $u_k > 0$ for all sufficiently big k, it follows that (3.3) holds for all such k. Suppose that (3.3) holds for all $k \geqslant k_1$. It will suffice to show that it holds also for $k_1 - 1$. Using (3.2) one obtains

$$(3.5) \quad \begin{aligned} u_n &= O(u_{n-1} + u_{n-2} + \cdots u_{n-n_1}) \\ &= O(u_{n-1+k_1} + u_{n-2+k_1+1} + \cdots u_{n-n_1+k_1+n_1-1}) \\ &= O(u_{n+k_1-1}), \end{aligned}$$

as had to be proved.

COROLLARY. *If $f_n = O(f_{n-1})$ then (3.4) holds.*

Proof. Suppose $f_{n+1} \leqslant \lambda f_n$ for $n \geqslant N_1$. Then, for $n \geqslant N_1$,

$$u_{n+1} = \sum_{k=1}^{N_1} f_k u_{n+1-k} + \sum_{k=N_1+1}^{n+1} f_k u_{n+1-k}$$

$$\leqslant \sum_{k=1}^{N_1} f_k u_{n+1-k} + \lambda \sum_{k=N_1+1}^{n+1} f_{k-1} u_{n+1-k}$$

$$\leqslant \sum_{k=1}^{N_1} f_k u_{n+1-k} + \lambda u_n$$

and Proposition 3.1 applies.

THEOREM 3.1. *A necessary and sufficient condition for $u_{n+1} \sim u_n$ is the existence of a sequence of non-negative numbers (a_n) such that $\sum_{n=1}^{\infty} a_n = 1$ satisfies the following two conditions:*

(3.6) $$\lim_{n \to \infty} \sum_{k=1}^{n} a_k \left(\frac{u_{n-k}}{u_n}\right) = 1.$$

(3.7) $$\lim_{N \to \infty} \limsup_{n \to \infty} \sum_{k=N+1}^{n} a_k \left(\frac{u_{n-k}}{u_n}\right) = 0.$$

Proof. Choosing $a_1 = a$, $a_n = 0$ for $n \geqslant 1$ shows the necessity. For the converse assertion consider first a sequence (a_n) satisfying the conditions of the theorem and also the assumption that 1 is the greatest devisor of $\{n : a_n > 0\}$. It will be shown that $u_{n+1} \sim u_n$.

It follows from (3.7) that for every $\epsilon > 0$ there exist integers N' and $N > N'$ such that $n \geqslant N$ implies

$$\sum_{k=N'+1}^{n} a_k u_{n-k} \leqslant \epsilon u_n$$

and by (3.6) there exists $N_1 > N$ such that $n > N_1$ implies

$$(1-\epsilon)u_n < \sum_{k=1}^{n} a_k u_{n-k} = \sum_{k=0}^{N'} a_k u_{n-k} + \sum_{k=N'+1}^{n} a_k u_{n-k}.$$

Therefore for $n > N_1$

$$(1-2\epsilon)u_n < \sum_{k=1}^{N'} a_k u_{n-k}.$$

Thus $u_n = 0(u_{n-1} + u_{n-2} + \cdots + u_{n-N'})$ and so $0 < m \leq M < \infty$ by Proposition 3.1. So a diagonal argument can be used to conclude that for any increasing sequence of positive integers there exists a subsequence (n') such that

(3.8) $$\lim \frac{u_{n'+i+1}}{u_{n'+i}} = h_i$$

exists for all i, and $m \leq h_i \leq M$. It follows from (3.6) that

$$u_{n+1} \sim a_1 u_n + \sum_{k=1}^{n} a_{k+1} u_{n-k}.$$

Divide through by u_n and set $n = n' + i$, where (n') is assumed to satisfy (3.8). It follows from (3.7) and the fact that $M < \infty$ that one may let n' tend to infinity to conclude

(3.9) $$h_i = a_1 + \sum_{k=1}^{\infty} \frac{a_{k+1}}{h_{i-k}h_{i-k+1}\cdots h_{i-1}}$$

$$= \sum_{k=1}^{\infty} \frac{a_k h_{i-k}}{h_{i-k}h_{i-k+1}\cdots h_{i-1}}.$$

Evidently there exists an increasing sequence satisfying (3.8) such that $h_0 = M$, and then a simple argument shows that (3.9) can be satisfied only if $h_i = M$ for all i. The argument will be needed once again and it is written out in the proof of Proposition 4.1 below. Observe that (3.9) is like (4.6) but with equality in place of inequality, a_k for f_k and 0 for λ. Finally, if h_i is identically equal to M and satisfies (3.9) then M must equal 1 and $u_{n+1} \sim u_n$ follow from Proposition 1.2. It only remains to remove the assumption that the greatest common devisor of $\{n : a_n > 0\}$ is 1. Suppose then it is d, $d > 1$. The sequence (a'_n) with $a'_n = a_{nd}$ will satisfy (3.6) and (3.7) with u'_n in place of u_n, where $u'_n = u_{nd}$. The argument then shows $u'_{n+1} \sim u'_n$ and $u_{n+1} \sim u_n$ follows by Proposition 1.2.

Remark. There is an alternative proof of Theorem 3.1 which is shorter and gives more insight, but which depends on a fact which seems to lie deeper. Having established $0 \leqslant m \leqslant M < \infty$ one can use a diagonal argument to conclude that for every increasing sequence of positive integers there exists a sub-sequence (n') such that $u_{n'-i}/u_{n'}$ tends to a finite limit g_i as n' tends to infinity, and from (3.6) and (3.7) one easily deduces

$$(3.10) \qquad g_i = \sum_{k=1}^{\infty} a_k \, g_{i+k}, \qquad g_0 = 1.$$

Now (3.10) says that g is a harmonic function for the Markov chain with state space $S = \{0, \pm 1, \pm 2, \ldots\}$ and transition probabilities $P_{j, j+k} = p_k$, where $p_k = a_k$ for $k > 0$, $p_k = 0$ for $k \leqslant 0$. The relevant fact, a proof of which can be found in [Doob, Snell, and Williamson, 1960], is that such Markov chains have no positive harmonic functions except constants. So $g(i) = 1$ for all i, which is the desired conclusion.

COROLLARY 1. *A necessary and sufficient condition for* $u_{n+1} \sim u_n$ *is*

(3.11) $$\lim_{N \to \infty} \limsup_{n \to \infty} \sum_{k=N}^{n} f_k \left(\frac{u_{n-k}}{u_n} \right) = 0.$$

Proof. Immediate from Theorem 3.1 and the renewal equation (1.7).

COROLLARY 2. *A sufficient condition for* $u_{n+1} \sim u_n$ *is*

(3.12) $$\sum_{k=N}^{\infty} (f_k/u_k) < \infty,$$

where N is such that $u_n > 0$ *for* $n \geq N$.

Proof. Use the previous corollary and note that by (1.9) $u_n \geq u_{n-k} u_k$.

4. *Conditions on the f_k implying* $u_{n+1} \sim u_n$. The assumptions made at the beginning of section 3 continue to be in force. Necessary and sufficient conditions on the f_k for $u_{n+1} \sim u_n$ are not known. Some sufficient conditions will be given however. One kind of condition that is known to be sufficient is that for large n, f_n be not too big relative to the preceding f_k's. In the simplest case one imposes only a condition on the ratio f_n/f_{n-1}. This is the only condition of this kind we will discuss here (see Theorem 4.1), though it has the major drawback that it is always violated when $f_n = 0$ for infinitely many n. A different kind of condition requires that the tails of the series $\Sigma k f_k$ go down sufficiently rapidly. Of course if these tails go down so rapidly that $\sum_{k=1}^{\infty} k f_k < \infty$ we know that u_n converges to a non-zero constant, so evidently $u_n \sim u_{n+1}$. It is somewhat surprising that weaker rate of decrease conditions on the tails also force $u_{n+1} \sim u_n$ (see Theorem 4.2).

We introduce the notation

(4.1) $$q_n = \frac{u_{n+1}}{u_n}, \qquad n \geq N_0,$$

where N_0 is the least N such that $u_n > 0$ for all $n \geq N$. Recall also the notations m, M introduced in Definition 3.1. When not otherwise indicated limits are taken as n tends to infinity.

PROPOSITION 4.1. $M \leq 1 \vee \lim\sup(f_{n+1}/f_n)$.

Proof. Suppose $\lim\sup(f_{n+1}/f_n) < \infty$, since otherwise there is nothing to prove. Then $0 < m \leq M \leq \infty$ by the corollary to Proposition 3.1. A diagonalization argument shows the existence of a sequence (n') of positive integers increasing to infinity so that

$$\lim_{n' \to \infty} q_{n'+i} = h_i$$

exists, $i = 0, \pm 1, \pm 2, \ldots$. Recall the renewal equation

(4.1) $$u_n = \sum_{k=1}^{n} f_k \, u_{n-k}, \; n = 1, 2, \ldots.$$

It implies that

$$1 \geq \sum_{k=1}^{N} f_k \left(\frac{u_{n-k}}{u_n}\right) = \sum_{k=1}^{N} \frac{f_k}{q_{n-1} q_{n-2} \cdots q_{n-k}}, \qquad n-N \geq N_0.$$

Put $n = n'+i$, and let n', then N, approach infinity to obtain

(4.2) $$1 \geq \sum_{k=1}^{\infty} \frac{f_k}{h_{i-1} h_{i-2} \cdots h_{i-k}}.$$

In place of (4.1) one may also write

(4.3) $$u_{n+1} = f_1 u_n + \sum_{k=1}^{n} f_{k+1} u_{n-k}, \quad n = 0, 1, \ldots.$$

Multiplying (4.1) by λ and subtracting from (4.3) leads to

(4.4) $$u_{n+1} = (f_1+\lambda)u_n + \sum_{k=1}^{n} (f_{k+1}-\lambda f_k)u_{n-k}, \quad n \geq 1.$$

Let $\lambda > \limsup(f_{n+1}/f_n)$. So there is a positive integer N_1 such that $(f_{k+1} - \lambda f_k) < 0$ for $k \geq N_1$. Therefore (4.4) implies

(4.5) $$q_n = \frac{u_{n+1}}{u_n} \leq (f_1 + \lambda) + \sum_{k=1}^{N} (f_{k+1} - \lambda f_k) \frac{u_{n-k}}{u_n}$$

$$= (f_1 + \lambda) + \sum_{k=1}^{N} \frac{(f_{k+1}-\lambda f_k)}{q_{n-1} q_{n-2} \cdots q_{n-k}}, \quad n \geq N \geq N_1 \vee N_0.$$

Again put $n = n'+i$, and let n' tend to infinity, obtaining

$$h_i \leq (f_1 + \lambda) + \sum_{k=1}^{N} \frac{(f_{k+1}-\lambda f_k)}{h_{i-1} h_{i-2} \cdots h_{i-k}}.$$

Now let N go to infinity and use (4.2) to justify

(4.6) $$h_i \leq (f_1 + \lambda) + \sum_{k=1}^{\infty} \frac{f_{k+1}}{h_{i-1} h_{i-2} \cdots h_{i-k}} - \lambda \sum_{k=1}^{\infty} \frac{f_k}{h_{i-1} h_{i-2} \cdots h_{i-k}}$$

$$= \lambda + \sum_{k=1}^{\infty} \frac{f_k}{h_{i-1} h_{i-2} \cdots h_{i-k}} (h_{i-k} - \lambda),$$

and (4.6) is equivalent to

(4.7) $$\sum_{k=1}^{\infty} \frac{f_k}{h_{i-1} h_{i-2} \cdots h_{i-k}} (h_i - h_{i-k}) \leq (\lambda - h_i)\left(1 - \sum_{k=1}^{\infty} \frac{f_k}{h_{i-1} h_{i-2} \cdots h_{i-k}}\right).$$

The sequence (n') can be chosen so that for some i $h_i \left(= \lim_{n' \to \infty} q_{n'+i}\right) = M$. Suppose $M > \lambda$. If $h_i = M$ the first factor on the right of (4.7) is negative, whilst the second factor is always non-negative by (4.2). On the left side of (4.7) each term is now non-negative. Thus both sides of (4.7) must vanish. In particular the second factor on the right must equal zero, and from the vanishing of the left side follows $M = h_i = h_{i-k}$ for each k such that $f_k > 0$. By iteration $h_j = M$ for all small enough j, say $j \leqslant j_0$ (of course this depends on the aperiodicity assumption (1.6)). So putting $i = j_0$ in (4.7) the second factor on the right will again vanish, that is $\sum_{k=1}^{\infty} (f_k/M^k) = 1$, proving $M = 1$. Since λ was an arbitrary number greater than $\limsup(f_{n+1}/f_n)$ the proposition is proved.

THEOREM 4.1. *If* $\limsup(f_{n+1}/f_n) \leqslant 1$ *then* $u_{n+1} \sim u_n$.

Proof. Follows from Proposition 4.1 and relation (3.1).

Write:

$$(4.8) \qquad r_n = \sum_{k=n+1}^{\infty} f_k, \qquad n = 0, 1, \ldots.$$

It will be shown that $u_{n+1} \sim u_n$ if r_n goes to zero fast enough.

Definition 4.1. Let n_0 be the least n such that $f_n > 0$. For $n > n_0$ let a_n be the positive number such that

$$(4.9) \qquad a_n f_1 + a_n^2 f + \ldots + a_n^n f_n = 1.$$

For $1 \leqslant n < n_0$ let $a_n = a_{n_0}$.

Note that (a_n) is a non-increasing sequence tending to 1.

PROPOSITION 4.2. *For all sufficiently big* n

$$(4.10) \qquad a_n \leqslant \exp[r_n / \sum_{k=1}^{n} k\, f_k] \leqslant \exp[r_n]$$

Proof. Let $n \geqslant n_0$, so that (4.9) holds. Since $a_n \geqslant 1$, $a_n^k = \exp[\log a_n^k] \geqslant 1 + \log a_n^k = 1 + k \log a_n$, and one obtains

$$1 = \sum_{n=1}^{n} a_n^k f_k \geqslant \sum_{k=1}^{n} f_k(1 + k \log a_n) = (1 - r_n) + \log a_n \sum_{k=1}^{n} k f_k,$$

and the first inequality in (4.10) follows. Since ultimately $\sum_{k=1}^{n} k f_k \geqslant 1$ the second inequality also will hold.

Recall that N_0 is the least N such that $u_n > 0$ for $n \geqslant N_0$.

PROPOSITION 4.3. *There exists a positive constant A such that*

(4.11) $\qquad (a_1 a_2 \ldots a_n) u_n \geqslant A, \quad n = N_0, N_0 + 1, \ldots.$

Proof. First it will be shown that there exists $B > 0$ such that

(4.12) $\qquad (a_1 a_2 \ldots a_m) u_{N_0+m} \geqslant B$

for all $m \geqslant 0$. The argument is by induction. Evidently B can be chosen so that (4.12) holds for $m = 0$. Suppose (4.12) holds for $0 \leqslant m < n$. Then, using the renewal equation,

$$(a_1 a_2 \ldots a_n) u_{N,0+n} \geqslant \sum_{k=1}^{n} f_k (a_1 a_2 \ldots a_{n-k}) u_{N_0+n-k} (a_{n-k+1} a_{n-k+2} \ldots a_n)$$

$$\geqslant B \sum_{k=1}^{n} f_k \, a_n^k = B,$$

so that (4.12) holds for all m. Evidently (4.11) holds with $A = B \, a_1^{N_0}$

PROPOSITION 4.4. *If* $\sum_{n=1}^{\infty} f_n (a_1 a_2 \ldots a_n) < \infty$ *then* $u_{n+1} \sim u_n$.

Proof. Use Proposition 4.3 and Corollary 2 of Proposition 3.1.

THEOREM 4.2. *If* $r_n = O(n^{-1})$ *then* $u_{n+1} \sim u_n$.

Proof. Put $\beta_n = (a_1 a_2 \ldots a_n)$, $n = 1, 2, \ldots, \beta_0 = 0$ and note

(4.13)
$$\sum_{n=1}^{\infty} f_n \beta_n = \sum_{n=1}^{\infty} f_n \sum_{j=1}^{n} (\beta_j - \beta_{j-1})$$

$$= \sum_{j=1}^{\infty} (\beta_j - \beta_{j-1}) \sum_{n=j}^{\infty} f_n$$

$$= \sum_{j=2}^{\infty} \beta_{j-1}(a_j - 1) r_{j-1} + \beta_1.$$

By (4.10) $a_j \leqslant \exp[r_j]$, and so since $r_j = O(j^{-1})$, $(a_j - 1) = O(j^{-1})$. To show that the series in the last member of (4.13) converges it remains to obtain an estimate on β_j. It may be assumed that $\sum_{k=1}^{\infty} k f_k$ diverges, since otherwise the conclusion of the theorem certainly holds. Then (4.10) shows that for any $\epsilon > 0$, $a_n \leqslant \exp[\epsilon r_n]$ for all sufficiently big n. By assumption there exists a positive constant c such that $r_n \leqslant c n^{-1}$ for all big enough n. So if $\epsilon = (2c)^{-1}$ one obtains $a_n \leqslant \exp[1/(2n)]$ and hence $\beta_n = O(n^{1/2})$ and the convergence of the series is assured.

Notes

CHAPTER 1. The approach to the ergodic theory of Markov chains presented here is due to Doblin [1937], [1940]. It is essentially Doblin's theory as completed during the quarter of a century following the publication of his papers that is presented here.

§ 0. The literature frequently refers to chains which are φ-recurrent for some φ as *recurrent in the sense* of Harris.

§ 1–3. The exposition here follows closely Doob [1953]; many of the basic ideas go back to Doblin [1937]. Some extensions of the theory due to Chung [1964], Jain–Jamison [1967], Jamison–Orey [1967], Orey [1959] are incorporated.

§ 4. The basic Proposition 4.2 is due to Blackwell [1955], and Theorem 4.1 is from Blackwell–Freedman [1964]. The other results are mostly easy consequences as was noted in Jamison–Orey [1967]. An example similar to Example 4.1 appears in Blackwell–Freedman [1964].

§ 5. Theorem 5.1 is taken from Jamison–Orey [1967]. For denumerable state space this result was proved in Blackwell–Freedman [1964]. The observation contained in Remark 5.1 was utilized by Harris–Robbins [1953].

§ 6. Uniform φ-recurrence is closely related to Condition (D) of Doob [1953]. The latter implies the existence of a φ such that S can be partitioned into a finite number of sets S_0, S_1, S_2, ..., S_n, where $\varphi(S_0) = 0$ and S_0 is inessential, while for $1 \leqslant i \leqslant n$ S_i is closed and the process on S_i is uniformly φ-recurrent. The importance of studying the process on D, with D a D-set was realized by Harris [1955], [1956]. A result like Theorem 6.2 was given in Orey [1959], but the statement of the theorem there contains some additional incorrect assertions.

§ 7. The result, Theorem 7.1 (iii), can be found in Doob [1953], Theorem 7.1 (i) is in Orey [1959]. The result and proof of Theorem 7.2 are due to Harris [1956]. Theorem 7.3 and its Corollary are from Jain [1966].

§ 8. The concept of a normal chain was introduced and investigated by Doblin [1940]. Theorem 8.1 and its Corollary are taken from Jain–Jamison [1967], and so is Theorem 8.2. Theorem 8.3 and the example of an anormal chain preceding it are due to Blackwell [1945]. More information about the anormal case is in Doblin [1940], Chung [1964], and Jain–Jamison [1967].

§ 9. The main result, Theorem 9.1, the propositions leading up to it, and the proofs, are due to Doblin [1940]. Jamison [1969] showed that Suslin's conjecture implies that Theorem 9.1 remains correct with the hypothesis of φ-non-singularity replaced by the assumption that there exists no uncountable class of disjoint closed sets. Recently a shorter proof of Theorem 9.1 was found by Harris, which is given in Proposition 51 of Chung [1964], with a correction in Harris [1969]. Proposition 9.1 was noted by Jain–Jamison [1967].

CHAPTER 2. For the denumerable case, results and historical references are in Chung [1960]. Proposition 1 (in more restricted form) was announced without proof by Harris [1955]; the proof given here is that of Levitan [1967]. Proposition 2 is a special case of the ergodic theorem of Chacon–Ornstein [1960], the value of the

limit following from Chacon [1962]. Theorem 1 was first proved by
Jain [1966]. The Example is a minor modification of one given by
Isaac [1967]. Theorem 3 is also due to Isaac [1967]. Theorem 4
appears in Duflo [1969], where further related results are also
proved. Metivier [1969] also contains results similar to Theorem 4.
Theorem 5 is new. For Theorem 6 and a discussion of its
implications see Duflo [1969], and in the denumerable case,
Kemeny–Snell–Knapp [1966]. The relevant results for random walk
are in Spitzer [1964] and Ornstein [1968].

CHAPTER 3. §1. A brief discussion of individual ratio limit
theorems in the case of denumerable state space is given in Chung
[1960], and there also an example, credited to F. J. Dyson, is given
to show that (DA) and (R) do not imply the existence of individual
ratio limits. Proposition 1.2 and Theorem 1.1 were stated in Orey
[1961]. The proof of Proposition 1.2 given here is due to Pruitt
[1965]. Theorem 1.2 is due to Jain [1969]; it improves results of
Levitan [1967]. Example 1.2 is discussed in Port [1965]. In
connection with Example 1.3 see Ornstein [1967] for random walk
on the line. Random walk on the integers is easier to deal with and
is covered by Theorem 2.1; in this context the strong ratio limit
property was established by Chung–Erdös [1951].

§2. Example 2.1 is taken from Orey [1961]. The integral
representation used is given in Kendall [1960]. In the important,
special case mentioned in the last sentence of the Example
$(SRLP)$ was originally established by Karlin–McGregor [1959].

Theorem 2.1 was proved by Kingman–Orey [1964]. It generalizes
a theorem of Chung–Erdös [1951]. Theorem 2.2 is due to
Folkman–Port [1966].

§3 and 4. The questions discussed here (and more general

ones) were first considered in De Bruijn–Erdös [1951], [1952], and [1953]. A necessary and sufficient condition for $u_{n+1} \sim u_n$ is given in these papers.

Proposition 3.1 is taken from Garsia [1963]. The necessary and sufficient condition of Theorem 3.1 is from Folkman–Port [1966]; another one is given in Garsia [1963]. Corollary 1 to Theorem 3.1 was established in Garsia–Orey–Rodemich [1962]; Corollary 2 goes back to De Bruijn and Erdös.

Theorem 4.1 is given in Garsia–Orey–Rodemich [1962]; there and in Garsia [1963] better results are also included. If the hypothesis of Theorem 4.1 is strengthened to $f_n \sim f_{n+1}$ the result can already be found in the papers of De Bruijn–Erdös. Proposition 4.1, 4.2 and Theorem 4.2 are due to Garsia [1963]. The proof given here is that of Garsia, incorporating a correction provided by Pruitt.

Bibliography

BLACKWELL, D., The existence of anormal chains, *Bull. Am. math. Soc.*, 465-68, (1945)

BLACKWELL, D., On transient Markov processes with a countable number of States and stationary transition probabilities, *Ann. math. Statist.*, **26**, 654-58, (1955)

BLACKWELL, D. and FREEDMAN, D., The tail σ-field of a Markov chain and a theorem of Orey, *Ann. math. Statist.*, **35**, 1291-95, (1964)

CHACON, R.V., Identification of the limit of operator averages, *J. Math. Mech.* II, 961-68, (1962)

CHACON, R.V. and ORNSTEIN, D., A general ergodic theorem, *Illinois J. Math.*, **4**, 153-60 (1960) Pre-print

CHUNG, K.L., *Markov chains with stationary transition probabilities*, Springer, Berlin, (1960)

CHUNG, K.L., The general theory of Markov processes according to Doblin, *Z. Wahrscheinlichkeitstheorie verw. Geb.* **2**, 230-54, (1964)

CHUNG, K.L. and ERDOS, P., Probability limit theorems assuming only the first moment, *Mem. Am. math. Soc.*, No. 6, 1-19, (1951)

DE BRUIJN, N.G. and ERDOS, P., On a recursion formula and some Tauberian theorems, *J. Res. natn. Bur. Stand.*, **50**, 161-64, (1953)

DE BRUIJN, N.G. and ERDOS, P., Some linear and some quadratic recursion formulas I, II, Koninkl. Ned. Akad., Wetenshap (A), **54**, 374-82 (1951); **55**, 152-63 (1952)

DOBLIN, W., Sur les proprietes asymptotiques de mouvement regis par certains types de chaines simples, *Bull. Math. Soc.* **39**, No.1, 57-115; No. 2, 3-61 (1937)

DOBLIN, W., Elements d'une theorie generale des chaines simple constants de Markoff, *Annls. scient. Éc. norm. sup.*, Paris, III Ser, **57**, 61–111, (1940)

DOOB, J.L., *Stochastic processes*, Wiley and Sons, New York (1953)

DOOB, J.L., A Markov chain theorem, *Probability and Statistics: The Harold Cramer Volume*. (edited by V. Grenander) Stockholm, 50–57 (1960)

DOOB, J.L., SNELL, J.L. and WILLIAMSON, R.E., Applications of boundary theory to sums of independent random variables, *Contributions to Probability and Statistics*, (edited by I. Olkin) Stanford Univ. Press (1960)

DUFLO, M., *Operateurs potentiels des chaines et des processus de Markov irreductibles*, (1969) Pre-print

FOGUEL, S.R., *The ergodic theory of Markov processes*, Van Nostrand, Princeton, N.J. (1969)

FOLKMAN, J.H. and PORT, S.C., On Markov chains with the strong ratio limit property, *J. Math. Mech.*, **15**, 113–21 (1966)

GARSIA, A., Some Tauberian theorems and the asymptotic behaviour of probabilities of recurrent events, *J. math. Analysis Applic.*, **7**, 146–62 (1963)

GARSIA, A., OREY, S., and RODEMICK, E., Asymptotic behaviour of successive coefficients of some power series, *Illinois J. Math.*, **6**, 620–29 (1962)

HARRIS, T.E., Recurrent Markov processes, II (abstract) *Ann. math. Statist.*, **26**, 152–53 (1955)

HARRIS, T.E. and ROBBINS, H., Ergodic theory of Markov chains admitting an infinite invariant measure, *Proc. natn. Acad. Sci. U.S.A.*, **39**, 860–64 (1953)

HARRIS, T.E., The existence of stationary measures for certain Markov processes, *Third Berkeley Symposium on Math. Stat. and Probability*, vol. II, 113–24 (1956)

HARRIS, T.E., Correction to a proof, *Z. Wahrscheinlichkeitstheorie verw. Geb.*, (1969)

ISAAC, R., On the ratio limit theorem for Markov processes recurrent in the sense of Harris, *Illinois J. Math*, **11**, 608–15 (1967)

JAIN, N., Some limit theorems for a general Markov process, *Z. Wahrscheinlishkeitstheorie verw. Geb.*, **6**, 206-23 (1966)

JAIN, N. and JAMISON, B., Contributions to Doeblin's theory of Markov processes, *Z. Wahrscheinlishkeitstheorie verw. Geb*, **8**, 19-40, (1967)

JAIN, N., The strong ratio limit property for some general Markov processes, *Ann. Math. Stat.*, **40**, 986-992 (1969)·

JAMISON, B., A result in Doblin's theory of Markov chains implied by Souslin's conjecture, (1969) Pre-print

JAMISON, B and OREY, S., Markov chains recurrent in the sense of Harris, *Z. Wahrscheinlishkeitstheorie verw. Geb.*, **8**, 41-48 (1967)

KARLIN, S and McGREGOR, J., Random Walks, *Illinois J. Math.*, **3**, 66-81 (1959)

KEMENY, J.G., SNELL, J.L., and KNAPP, A.W., *Denumerable Markov chains*, Van Nostrand, Princeton, N.J. (1966)

KENDALL, D,G., Unitary dilutions of Markov transition operators, and the corresponding integral representation for transition-probability matrixes, *Probability and Statistics: The Harold Cramer Volume*, (edited by V. Grenander) Stockholm, 138-61 (1960)

KINGMAN, J.F.C. and OREY, S., Ratio limit theorems for Markov chains, *Proc. Am. math. Soc.*, **15**, 907-10 (1964)

LEVITAN, M., *Some ratio limit theorems for a general state space Markov Process*, Thesis, University of Minnesota (1967)

LOEVE, M., *Probability theory*, 2nd Ed., Van Nostrand, Princeton, N.J. (1955)

METIVIER, M., Existence of an invariant measure and an Ornstein's ergodic theorem, *Ann. Math. Stat.*, **40**, 74-96 (1969)

NEVEU, J., *Mathematical foundations of the calculus of probability*, Holden Day, San Francisco (1965)

OREY, S., Recurrent Markov chains, *Pacif. J. Math.*, **9**, 805-27 (1959)

OREY, S., Strong ratio limit property, *Bull. Am. math. Soc.*, **67**, 571-74 (1961)

ORNSTEIN, D., A limit theorem for independent random variables, *Fifth Berkeley Symposium on Probability and Statistics, California, 1965-6*, Vol. II, part 2, 213-16, Univ. of California Press, Berkeley, Cal. (1962)

ORNSTEIN, D., Random Walks I, II, Trans. *Am. math. Soc.*, **138**, 1-43 (1969); ibid, **138**, 45-60 (1969)

PORT, S.C., On random walks with a reflecting barrier, *Trans. Am. math. Soc.*, **5**, 362-70 (1965)

PRUITT, W., Strong ratio limit property for R-recurrent Markov chains, *Proc. Am. math. Soc.*, **16**, 196-200 (1965)

SPITZER, F., *Principles of random walk*, Van Nostrand, Princeton, N.J. (1964)

INDEX

(A), 66
Adapted, 2
a.e., 6
Anormal, 36
Aperiodic, 15
a.s., 3

$\mathcal{B}_A$, 29

Closed, 4
C-*set*, 7
Cycle, 13

(D), 65
(DR), 66
D-*set*, 29
$D^n(\mu, A; \nu, B)$, 57

E_μ, E_x, 2
E^0, E^∞, 36
Essential, 12

f_k, 66

Harmonic, 9

Improperly essential, 12
Indecomposable, 4
Indicator function, 2
Initial distribution, 1
Individual ratios, 65

Inessential, 12
Invariant event, random variable, σ-field, 15
Invariant (probability) measure, 30
Irreducible, 4

Jointly measurable, 15

$L(x, B)$, 3

Markov chain, 3
Markov property, 3
Maximal indecomposable, 42
Measure, 3

Non-singular, 42
Normal, 36

P_A, 28
$_B P^m(x, A)$, 26
P_x, 2
P_μ, 1
P-invariant event, random variable, σ-field, 15
P-tail event, random variable, σ-field, 16
Period(ic), 15
Properly essential, 12

INDEX

(R), 65
$R^n(\mu,A;\nu,B)$, $R^n(x,A;y,B)$, 49
Random walk, 78
Recurrent, 4
Renewal equation, 67
Reversible, 83

Separable, 5
Shift operator, 2
Space time process, 16
Space time harmonic, 16
Stochastic process, 2
Strong ratio limit property ($SRLP$), 70

Tail event, random variable, σ-field, 15
Transition probabilities, 1
Transition probability density, 5
Trivial σ-field, 16

Uniformly recurrent, 26

ϵ_x, 2
φ-irreducible etc., see irreducible etc.
μP, $||\mu||$, 19
θ, 2

VAN NOSTRAND REINHOLD MATHEMATICAL STUDIES are paperbacks focusing on the living and growing aspects of mathematics. They are not reprints, but original publications. They are intended to provide a setting for experimental, heuristic and informal writing in mathematics that may be research or may be exposition.

Under the editorship of Paul R. Halmos and Frederick W. Gehring, lecture notes, trial manuscripts, and other informal mathematical studies will be published in inexpensive, paperback format.

P. R. HALMOS received his Ph.D. from the University of Illinois, and spent three years at the Institute for Advanced Study, two of them as Assistant to John von Nuemann. He taught at the Universities of Chicago, Michigan, and Hawaii and is presently Professor of Mathematics at Indiana University.

F. W. GEHRING received his Ph.D. from Cambridge University, England. He has held Visiting Professorships at Harvard and Stanford Universities as well as Guggenheim, Fulbright, and NSF Fellowships at the University of Helsinki and the Eidgenössische Technische Hochschule in Zürich. He is presently Professor of Mathematics at the University of Michigan.

A VAN NOSTRAND REINHOLD MATHEMATICAL STUDY
under the general editorship of
PAUL R. HALMOS
Indiana University
FREDERICK W. GEHRING
The University of Michigan

About this book:

These notes are based on a course given in 1968, first at the University of Minnesota then at Westfield College, University of London. They deal with topics in the ergodic theory of discrete time Markov chains with stationary transition probabilities and arbitrary measurable space for state space. The first chapter is essentially a presentation of Doblin's theory as completed during the quarter of a century following the publication of his papers. Related ergodic problems are developed in the final two chapters. The treatment is probabilistic and the necessary concepts from probability theory are developed, making the notes largely self-contained.

The Author:
STEVEN OREY (Ph.D. Cornell University) is Professor of Mathematics at the University of Minnesota. He has published numerous articles in mathematical journals in the areas of probability theory and logic.

VAN NOSTRAND REINHOLD COMPANY
Windsor House, 46 Victoria Street,
London, S.W.1.   442 06299 0